ESSENTIAL FORMULAE FOR ELECTRONIC AND ELECTRICAL ENGINEERS

Second Edition

Noel M. Morris

MACMILLAN

First published 1974 (A4 format) by
THE MACMILLAN PRESS LTD
Houndmills, Basingstoke, Hampshire RG21 2XS
and London
Companies and representatives
throughout the world

ISBN 0–333–59474–6

A catalogue record for this book is available from the British Library.

Published in pocket-book format 1983
Second edition 1993
Reprinted 1994

Printed in Hong Kong

CONTENTS

iii

ix

xiii

PREFACE

As with earlier editions, the book contains essential formulae in the fields of electronic and electrical engineering, measurements and control, logic, telecommunications and mathematics. This edition reflects changes in technology and the general expansion in the fields of study listed above. It includes changes in nomenclature, new equations, and new sections on mensuration, hyperbolic formulae, matrices and z-transforms.

The book will be of value not only to students following a wide range of courses including science-based studies at school and college, BTEC courses (National and Higher National Certificate and Diploma) and B.Sc. courses, but also to established engineers who use formulae in their work.

Noel M. Morris

RESISTOR COLOUR CODE

(See Figure 1)

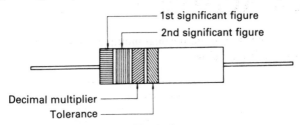

Figure 1

Colour	Significant figure	Decimal multiplier	Tolerance (per cent)
silver		0.01	10
gold		0.1	5
black	0	1	
brown	1	10	
red	2	10^2	
orange	3	10^3	
yellow	4	10^4	
green	5	10^5	
blue	6	10^6	
violet	7	10^7	
grey	8	10^8	
white	9	10^9	

A useful mnemonic to remember the numerical colour sequence is:

Bye **B**ye **R**osie, **O**ff **Y**ou **G**o, **B**ristol **V**ia **G**reat **W**estern

PREFERRED VALUES FOR RESISTORS AND CAPACITORS

The values in use are decimal multiples and submultiples of those listed below.

Percentage tolerance		
20%	10%	5%
10	10	10
		11
	12	12
		13
15	15	15
		16
	18	18
		20
22	22	22
		24
	27	27
		30
33	33	33
		36
	39	39
		43
47	47	47
		51
	56	56
		62
68	68	68
		75
	82	82
		91

MULTIPLES AND SUBMULTIPLES OF 10

Symbol	Prefix	Multiple
T	tera	10^{12}
G	giga	10^{9}
M	mega	10^{6}
k	kilo	10^{3}
c	centi	10^{-2}
m	milli	10^{-3}
μ	micro	10^{-6}
n	nano	10^{-9}
p	pico	10^{-12}
f	femto	10^{-15}
a	atto	10^{-18}

GREEK ALPHABET

Name	Capital	Lower case	Name	Capital	Lower case
alpha	A	α	nu	N	ν
beta	B	β	xi	Ξ	ξ
gamma	Γ	γ	omicron	O	o
delta	Δ	δ	pi	Π	π
epsilon	E	ϵ	rho	P	ρ
zeta	Z	ζ	sigma	Σ	σ
eta	H	η	tau	T	τ
theta	Θ	θ	upsilon	Y	υ
iota	I	ι	phi	Φ	ϕ
kappa	K	κ	chi	X	χ
lambda	Λ	λ	psi	Ψ	ψ
mu	M	μ	omega	Ω	ω

DIMENSIONS AND DIMENSIONAL ANALYSIS

Derived units in the MLTQ system

Quantity	Symbol	Unit-symbol	Dimension
Energy	W	J	ML^2T^{-2}
Power	P	W	ML^2T^{-3}
Charge	Q	C	Q
Current	I	A	$T^{-1}Q$
e.m.f.	E	V	$ML^2T^{-2}Q^{-1}$
Resistance	R		
Reactance	X	Ω	$ML^2T^{-1}Q^{-2}$
Impedance	Z		
Inductance	L	H	ML^2Q^{-2}
Magnetic flux	Φ	Wb	$ML^2T^{-1}Q^{-1}$
Magnetic flux density	B	T	$MT^{-1}Q^{-1}$
Magnetic field intensity	H	A/m	$L^{-1}T^{-1}Q$
Capacitance	C	F	$M^{-1}L^{-2}T^2Q^2$
Permeability	μ	H/m	MLQ^{-2}
Permittivity	ϵ	F/m	$M^{-1}L^{-3}T^2Q^2$
Electric flux	Q	C	Q
Electric flux density	D	C/m^2	$L^{-2}Q$
Electric field intensity	E	V/m	$MLT^{-2}Q^{-1}$

Basic and supplementary SI units

Quantity	Unit	Unit-symbol
Length, L	metre	m
Mass, M	kilogram	kg
Time, T	second	s
Current, I	ampere	A
Temperature	Kelvin	K
Luminous intensity	candela	cd
Plane angle	radian	rad
Solid angle	steradian	sr

MECHANICS

a = acceleration

F = force

J = moment of inertia

m = mass

r = radius

s = distance

t = time

T = torque

u = initial velocity

v = final velocity

α = angular acceleration

θ = angular displacement

ω = angular velocity

Equations of motion

$$v = u + at \qquad \text{m/s}$$

$$s = ut + \frac{1}{2}at^2 \qquad \text{m}$$

$$v^2 = u^2 + 2as \qquad (\text{m/s})^2$$

$$s = \theta r \qquad \text{m}$$

$$v = \omega r \qquad \text{m/s}$$

$$a = \alpha r \qquad \text{m/s}^2$$

Torque

$$T = Fr = J\alpha \qquad \text{Nm}$$

Force, work, energy, and power

$$\text{Force} = F = ma \qquad \text{N}$$

$$\text{Work} = W = Fs = T\theta \quad \text{J}$$

$$\text{Kinetic energy} = T = \frac{1}{2}mv^2 = \frac{1}{2}J\omega^2 \qquad \text{J}$$

$$\text{Power} = P = \omega T \qquad \text{W}$$

ELECTROSTATICS

A = area
C = capacitance
c_0 = speed of an e.m. wave
d = distance
D = electric flux density
e = charge on an electron
E = electric field strength
F = force in newtons
h = height
l = length

m_e = electron rest mass
m_n = neutron rest mass
m_p = proton rest mass
n = integer
Q = charge
r = radius
v = velocity
Z = impedance
ϵ = permittivity

6

Data relating to the electron, the proton, the neutron, and free space

Electron

      charge $e = -1.602 \times 10^{-19}$ C

      rest mass $m_e = 9.109 \times 10^{-31}$ kg

      charge-to-mass ratio $e/m_e = 1.759 \times 10^{11}$ C/kg

Proton

      charge $e = 1.602 \times 10^{-19}$ C

      rest mass $m_p = 1.673 \times 10^{-27}$ kg

      charge-to-mass ratio $e/m_p = 0.906 \times 10^{8}$ C/kg

Neutron

      rest mass $m_n = 1.675 \times 10^{-27}$ kg

Free space

      electric constant $\epsilon_0 = 8.854 \times 10^{-12}$ F/m

      intrinsic impedance $Z_0 = 376.9$ Ω

      speed of e.m. waves $c_0 = 2.998 \times 10^{8}$ m/s

Electric flux density $D = Q/A$ C/m^2

Electric field strength $E = \dfrac{Q}{4\pi\epsilon d^2}$ N/C or V/m

Absolute permittivity $\epsilon = \epsilon_0\epsilon_r = D/E$ F/m

Force between two charges $F = \dfrac{Q_1 Q_2}{4\pi\epsilon d^2}$ N

Energy stored in a cubic metre of dielectric

$$W = \frac{1}{2} \epsilon E^2 = \frac{1}{2} DE = \frac{1}{2} D^2/\epsilon \qquad \text{J}$$

Capacitance $\quad C = Q/V$ $\qquad$ F

Energy stored in a capacitor $\quad W = \frac{1}{2} CV^2$ $\qquad$ J

Capacitance of a parallel-plate capacitor with n plates

$$C = \frac{\epsilon(n-1)a}{d} \qquad \text{F}$$

Capacitors in parallel $\quad C = C_1 + C_2 + \ldots$

Capacitors in series $\quad \dfrac{1}{C} = \dfrac{1}{C_1} + \dfrac{1}{C_2} + \ldots$

Ratio of field strengths in two dielectrics in series Two dielectrics A and B in series, having the same area: the ratio of the field strengths in dielectrics A and B of permittivities ϵ_A and ϵ_B is

$$\frac{E_A}{E_B} = \frac{\epsilon_A}{\epsilon_B}$$

Capacitance of concentric spheres $C = \dfrac{4\pi\epsilon}{\dfrac{1}{r_1} - \dfrac{1}{r_2}}$ F

Capacitance of an isolated sphere $C = 4\pi r \epsilon$ F

Capacitance per metre of two concentric cylinders

$$C = \frac{2\pi\epsilon}{\log_e r_2/r_1} \qquad \text{F}$$

Capacitance per metre of two parallel bare wires

$$C = \frac{\pi\epsilon}{\log_e d/r} \qquad \text{F}$$

Capacitance per metre of a single conductor to earth

$$C_1 = \frac{2\pi\epsilon}{\log_e 2h/r} \qquad \text{F}$$

Capacitance per metre of line 1 of an isolated three-phase line

$$C_1 = \frac{2\pi\epsilon}{\log_e (d_{13}d_{12}/r\, d_{23})} \qquad \text{F}$$

9

Force on an isolated electron in an electric field $F = Ee$ N

Final velocity of a free electron in an electric field

$$v = (2V \ e/m)^{\frac{1}{2}} \qquad \text{m/s}$$

Transverse deflection of an electron in an electrostatic C.R.T.

$$\text{deflection} = \frac{1}{2} \frac{e}{m} \frac{V}{d} \left(\frac{l}{v} \right)^2 \qquad \text{m}$$

Force of attraction between charged plates $F = \frac{1}{2} \epsilon A E^2$ N

ELECTROMAGNETISM

a = area
B = flux density
d = distance
e = charge on an electron
E = e.m.f.
f = frequency
F = force in newtons
F = m.m.f.
H = magnetic field intensity
I = current
k = constant
k = coupling coefficient
l = length
m = mass of an electron

M = mutual inductance
N = number of turns
r = radius
S = reluctance
T = torque
u = velocity
v = velocity
v = volume
W = energy
x = length
θ = angle
μ = permeability
Φ = flux

Magnetic constant of free space $\mu_0 = 4\pi \times 10^{-7}$ H/m

Reluctance $S = l/\mu a$ A/Wb

Magnetomotive force $F = NI$ A or At

Magnetic flux $\Phi = F/S$ Wb

Magnetic flux in a composite circuit $\Phi = \dfrac{F}{S_1 + S_2 + \ldots}$ Wb

Leakage factor   total flux/useful flux

Magnetic flux density $B = \dfrac{\Phi}{a}$ T

Magnetic field intensity $H = \dfrac{NI}{l} = \dfrac{F}{l}$ A/m or A t/m

Absolute permeability $\mu = \mu_0 \mu_r = B/H$ H/m

Force on a current-carrying conductor in a magnetic field, the conductor being at angle θ to the field

$$F = BlI \sin \theta \qquad \text{N}$$

the force being perpendicular to both the field and the current.

Torque on a coil in a magnetic field $T = BaNI \sin \theta$ Nm
where θ is the angle between the direction of the magnetic field and a line perpendicular to the axis of the coil.

Force on an isolated electron moving at velocity u in a magnetic field $F = Beu$
                                                                          N

the force being perpendicular to both B and u

Deflection of an electron in a magnetic field

$$\sin \theta = B \times \frac{e}{m} \times \frac{l}{u}$$

E.M.F. generated in a conductor $E = d\Phi/dt = Blv$ V

E.M.F. induced in a coil $E = \dfrac{d}{dt}(N\Phi)$ V

Magnetising force in a long solenoid $H = \dfrac{NI}{l}$ A/m

Magnetising force at distance x from the centre of a straight wire

$$H = \frac{I}{2\pi x} \qquad \text{A/m}$$

Flux density in air at distance x from the centre of a straight wire

$$B = \frac{I\mu_0}{2\pi x} \qquad \text{T}$$

Energy stored per cubic metre in a magnetic field

$$W = \frac{B^2}{2\mu} = \frac{1}{2}\mu H^2 \qquad \text{J}$$

Force between two magnetised surfaces $F = \dfrac{B^2 a}{2\mu_0}$ N

Force per metre between two parallel conductors

$$F = \frac{\mu_0 I_1 I_2}{2\pi d} \qquad \text{N}$$

Inductance $L = N\Phi/I = N\, d\Phi/di$ H

Inductance of a homogeneous magnetic circuit

$$L = \frac{N^2 \mu a}{l}$$ H

Inductance per metre of two parallel wires d metres apart and each of radius r

$$L = \frac{\mu_0}{\pi} \log_e \frac{d}{r}$$ H

Inductance per metre of a concentric cable

$$L = \frac{\mu_0}{\pi} \log_e \frac{r_1}{r_2}$$ H

Magnetic energy stored in an inductor $W = \frac{1}{2} L I^2$ J

Self induced e.m.f. $E = L\, di/dt$ V

Inductance of coils in series, not magnetically coupled

$$L = L_1 + L_2 + \ldots$$

Inductance of coils in parallel, not magnetically coupled

$$\frac{1}{L} = \frac{1}{L_1} + \frac{1}{L_2} + \dots$$

Mutual inductance $\quad M = \dfrac{N_2 \Phi_{12}}{I_1}$ $\qquad$ H

where $N_2 \Phi_{12}$ are the flux linkages associated with N_2 when current I_1 flows in the primary coil.

Coupling coefficient $\quad k = M/(L_1 L_2)^{1/2}$

Inductance of two coils in series which are magnetically coupled $\quad L = L_1 + L_2 \pm 2M$ $\qquad$ H

Coupled circuits with alternating e.m.f. E_1 **in the primary and** E_2 **in the secondary**

$$E_1 = (R_1 + j\omega L_1)I_1 \pm j\omega M I_2$$
$$E_2 = \pm j\omega M I_1 + (R_2 + j\omega L_2)I_2$$

Hysteresis loss $\quad kvf\,(B_{max})^n$ $\qquad$ W

where $\quad 1.5 < n < 2.5$

ELECTRIC CIRCUITS

a = area
E = e.m.f.
I = current
l = length
P = power
Q = charge
R = resistance
t = time
Y = admittance
α_0 = resistance-temperature coefficient at 0°C
ρ = resistivity
θ = temperature

Resistance $\quad R = \dfrac{\rho l}{a} \qquad\qquad\qquad\qquad \Omega$

Resistors in series $\quad R = R_1 + R_2 + \ldots$

Resistors in parallel $\quad \dfrac{1}{R} = \dfrac{1}{R_1} + \dfrac{1}{R_2} + \ldots$

Variation of resistance with temperature

$$R = R_0 (1 + \alpha\theta + \beta\theta^2)$$

$$\frac{R_2}{R_1} = \frac{1 + \alpha_0\theta_2}{1 + \alpha_0\theta_1}$$

$$\alpha_n = \frac{1}{\theta_n + 1/\alpha_0}$$

Conductance $G = 1/R$ S

Voltage $E = IR$ V

Current $I = \dfrac{Q}{t}$ A

Power $P = EI = I^2R = \dfrac{E^2}{R}$ W

Energy $W = EIt = I^2Rt = \dfrac{E^2t}{R}$ J

Division of current in parallel resistors R_1 and R_2

$$\text{Current in } R_1 = \text{total current} \times \frac{R_2}{R_1 + R_2}$$

Division of voltage between series resistors R_1 and R_2

$$\text{Voltage across } R_1 = \text{total voltage} \times \frac{R_1}{R_1 + R_2}$$

CIRCUIT THEOREMS

Kirchhoff's laws   *Current (KCL) law*. The total current flowing into any node in a circuit is equal to the total current flowing away from that node. In general, $\Sigma I = 0$ at any node (see Figure 2(a)).

Voltage (KVL) law. In any closed circuit, the algebraic sum of the potential drops is equal to the resultant e.m.f. in the loop. In general, $\Sigma E = \Sigma IR$ (see Figure 2(b)).

Thévenin's theorem   A two-terminal network can be replaced by a voltage-source equivalent electrical network having an e.m.f. E and an internal impedance Z (see Figure 3). E.m.f. E is the no-load voltage which appears between the two terminals, and Z is the impedance of the network measured between these terminals with the load disconnected and E meanwhile being replaced by its internal impedance.

Norton's theorem   A two-terminal network can be replaced by a current—source equivalent network having an internal current source I shunted by an admittance Y (see Figure 4). Current I is the current which flows in a short circuit applied between the terminals of the network. Admittance Y is the admittance measured between the network terminals with the load disconnected, and with the current generator being replaced by its internal admittance.

(a)

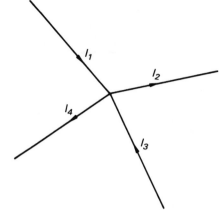

Kirchhoff's first law
$I_1 - I_2 + I_3 - I_4 = 0$

(b)

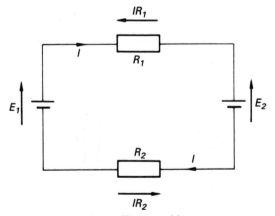

Kirchhoff's second law
$E_1 - E_2 = IR_1 + IR_2$

Figure 2

19

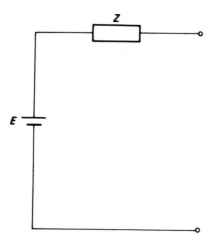

Figure 3

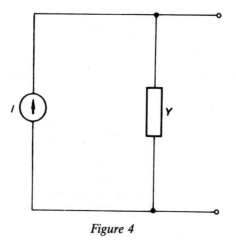

Figure 4

20

Relationship between Thévenin's and Norton's theorems

$$Z = 1/Y \quad E = I/Y = IZ$$

Millman's theorem (See Figure 5). If any number of admittances Y_1, Y_2, Y_3, etc., meet at a common point O', and the e.m.fs from another point O to the free ends of these impedances are E_1, E_2, E_3, etc., then the voltage between points O and O' is

$$E_{O'O} = \frac{E_{10}Y_1 + E_{20}Y_2 + \ldots}{Y_1 + Y_2 + \ldots}$$

In general

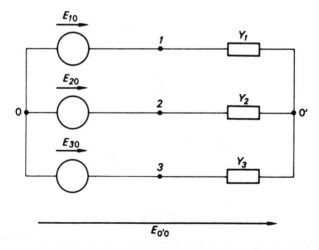

Figure 5

21

$$E_{O'O} = \frac{\sum\limits_{k=1}^{k=n} E_{k0}Y_k}{\sum\limits_{k=1}^{k=n} Y_k}$$

Star—mesh transformation   The star circuit in Figure 6 is transformed into its mesh equivalent by the following relationships

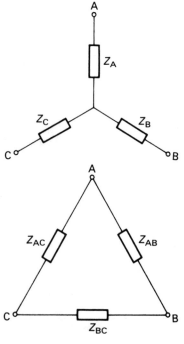

Figure 6

22

$$Z_{AB} = Z_A + Z_B + \frac{Z_A Z_B}{Z_C} \quad \text{or} \quad Y_{AB} = \frac{Y_A Y_B}{Y_A + Y_B + Y_C}$$

$$Z_{BC} = Z_B + Z_C + \frac{Z_B Z_C}{Z_A} \quad \text{or} \quad Y_{BC} = \frac{Y_B Y_C}{Y_A + Y_B + Y_C}$$

$$Z_{AC} = Z_A + Z_C + \frac{Z_A Z_C}{Z_B} \quad \text{or} \quad Y_{AC} = \frac{Y_A Y_C}{Y_A + Y_B + Y_C}$$

In general, if the star network has Q terminals, then

$$Y_{PQ} = \frac{Y_P Y_Q}{\displaystyle\sum_{k=A}^{k=Q} Y_k}$$

The general theorem is known as Rosen's theorem.

Delta—star transformation　　In Figure 6, the delta network is transformed into its star equivalent by the following

$$Z_A = \frac{Z_{AC} Z_{AB}}{Z_{AB} + Z_{BC} + Z_{AC}} \quad \text{or} \quad Y_A = Y_{AC} + Y_{AB} + \frac{Y_{AC} Y_{AB}}{Y_{BC}}$$

$$Z_B = \frac{Z_{AB} Z_{BC}}{Z_{AB} + Z_{BC} + Z_{AC}} \quad \text{or} \quad Y_B = Y_{AB} + Y_{BC} + \frac{Y_{AB} Y_{BC}}{Y_{AC}}$$

$$Z_C = \frac{Z_{AC} Z_{BC}}{Z_{AB} + Z_{BC} + Z_{AC}} \quad \text{or} \quad Y_C = Y_{AC} + Y_{BC} + \frac{Y_{AC} Y_{BC}}{Y_{AB}}$$

Maximum power transfer theorem

For any system
A pure resistance load will abstract maximum power from
a network when the resistance of the load is equal to the
internal resistance or output resistance of the network.

For an a.c. system

(a) If the load comprises a fixed reactance in series with a
    variable resistance, maximum power is transferred
    when the resistance of the load is equal to the sum of
    the magnitude of the internal impedance of the active
    network and the reactance of the load.
(b) If the load has a constant power factor but a variable
    impedance, maximum power is transferred when the
    magnitude of the load impedance is equal to the mag-
    nitude of the internal impedance of the source.
(c) If the load resistance and the load reactance are inde-
    pendently variable, maximum power is transferred
    when the load impedance is equal to the complex
    conjugate of the internal impedance of the source.

Superposition theorem   The current in each branch of a
network is the sum of currents in that branch due to each
e.m.f. acting alone, other e.m.fs being replaced meanwhile by their internal impedances.

Reciprocity theorem   If an e.m.f. E acting in one branch
of a network causes current I to flow in a second branch,

then the same e.m.f. acting in the second branch produces the same current in the first branch.

Compensation theorem   If the impedance of any branch of a network in which the current is I is changed by δZ, then the changes of current at all points in the network may be calculated on the assumption that an e.m.f. $-I\delta Z$ has been introduced into the changed branch. When calculating the change in current, all other e.m.fs are replaced by their internal impedances.

COMPLEX NUMBERS

Rectangular form j^2 (or i^2) $= -1$

The number $a + jb$ is described as a complex number, the term jb being called the 'imaginary' part or quadrature part of the number (see Figure 7). The *complex conjugate* of $a + jb$ is $a - jb$. The quantities a and b above are perpendicular to one another, and are described as the *rectangular coordinates* of the quantity.

$$(a + jb) \pm (c + jd) = (a \pm c) + j(b \pm d)$$

$$(a + jb) \times (c + jd) = (ac - bd) + j(ad + bc)$$

$$\frac{1}{a + jb} = \frac{a - jb}{a^2 + b^2}$$

$$\frac{a + jd}{c + jd} = \frac{(a + jb)(c + jd)}{c^2 + d^2} = \frac{(ac + bd) + j(bc - ad)}{c^2 + d^2}$$

25

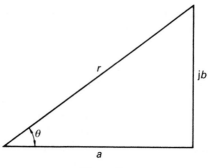

Figure 7

Polar form

$$r\underline{/\theta} = r(\cos\theta + j\sin\theta) = a + jb$$

where $r = (a^2 + b^2)^{\frac{1}{2}}$

and $\theta = \tan^{-1}\dfrac{b}{a}$

$$r_1\underline{/\theta_1} \times r_2\underline{/\theta_2} = r_1r_2\underline{/\theta_1 + \theta_2}$$

$$r_1\underline{/\theta_1} \div r_2\underline{/\theta_2} = \frac{r_1}{r_2}\underline{/\theta_1 - \theta_2}$$

Exponential form

$$re^{j\theta} = r(\cos\theta + j\sin\theta)$$

$$r_1e^{j\theta_1} \times r_2\,e^{j\theta_2} = r_1r_2\,e^{j(\theta_1 + \theta_2)}$$

$$(r_1\,e^{j\theta_1})/(r_2\,e^{j\theta_2}) = \frac{r_1}{r_2}\,e^{j(\theta_1 - \theta_2)}$$

26

Complex conjugate   The complex conjugate of $r\underline{/\theta}$ is $r\underline{/-\theta}$, and the complex conjugate of $(a + \mathrm{j}b)$ is $(a - \mathrm{j}b)$, so that

$$r\underline{/\theta} \times r\underline{/-\theta} = r^2$$

and

$$(a + \mathrm{j}b) \times (a - \mathrm{j}b) = (a^2 + b^2)$$

Powers and roots of a complex number   The nth power of a complex number is

$$(r\underline{/\theta})^n = r^n\underline{/n\theta}$$

There are n roots of the complex number $r\underline{/\theta}$ as follows

$$(r\underline{/\theta})^{1/n} = r^{1/n}\underline{/\dfrac{\theta + [n \times 360]}{n}}$$

where n has a value in the range $0, 1, 2 \ldots (n - 1)$.

De Moivre's theorem   For all rational values of n $\cos n\theta + \mathrm{j} \sin n\theta$ is one of the values of $(\cos \theta + \mathrm{j} \sin \theta)^n$

SINGLE PHASE A.C. CIRCUITS

B = susceptance
C = capacitance
e = instantaneous value of e.m.f.
E = r.m.s. value of e.m.f.
E_{av} = average value of e.m.f.
E_m = maximum value of e.m.f.
f = frequency
f_0 = oscillatory frequency
G = conductance
I = current
L = inductance
Q = Q-factor
R = resistance
T = periodic time
X = reactance
Y = admittance
ϕ = phase angle
ω = angular frequency
ω_0 = oscillatory angular frequency

Frequency $f = 1 \text{ (periodic time)} = 1/T$ Hz

Angular frequency $\omega = 2\pi f$ rad/s

Instantaneous e.m.f. $e = E_m \sin \omega t$

Average e.m.f. Average value for n equidistant mid-ordinates over a half cycle (see Fig. 8)

$$\frac{e_1 + e_2 + \cdots + e_n}{n}$$

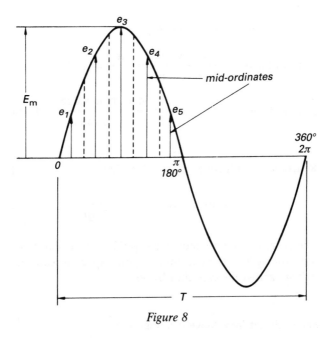

Figure 8

R.M.S. value $E = \{\text{average value of } (e)^2\}^{1/2}$
$= \{(e_1^2 + e_2^2 + \cdots + e_n^2)/n\}^{1/2}$

Average value of a rectified sine wave

$$E_{av} = \frac{2}{\pi} E_m = 0.637 E_m$$

Sine wave

r.m.s. value $= E = 0.707 E_m$

form factor $= \dfrac{\text{r.m.s. value}}{\text{average value}} = \dfrac{\pi}{2\sqrt{2}} = 1.111$

peak factor or crest factor $= \dfrac{\text{maximum value}}{\text{r.m.s. value}} = 1.414$

Reactance Inductive: $X_{L} = \omega L = 2\pi f L$ Ω

Capacitive: $X_c = \dfrac{1}{\omega C} = \dfrac{1}{2\pi f C}$ Ω

The current in a pure inductor lags the voltage across the inductor by 90°. The current in a pure capacitor leads the voltage across the capacitor by 90°.

Series circuit impedance (see Figure 9)

$$Z = R + jX = |Z| \underline{/\phi}$$

where

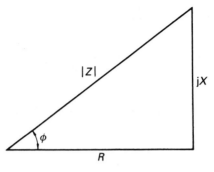

Figure 9

$$|Z| = \{R^2 + (X_L - X_C)^2\}^{1/2} = \left\{ R^2 + \left(\omega L - \frac{1}{\omega C} \right)^2 \right\}^{1/2}$$

and $\quad \phi = \tan^{-1} \dfrac{X}{R} = \tan^{-1} \dfrac{(X_L - X_C)}{R}$

Phase angle: if $X_L > X_C$, then I lags behind V.
if $X_L > X_C$, then I leads V.

Impedances in series

$$Z = (R_1 + R_2 + \cdots) + j(X_1 + X_2 + \cdots)$$
$$= R_T + jX_T = |Z| \underline{/\phi}$$

where $\quad |Z| = (R_T^2 + X_T^2)^{1/2} \quad$ and $\quad \phi = \tan^{-1} \dfrac{X_T}{R_T}$

31

Current in a series circuit

$$I = \frac{V}{Z} = \frac{V}{R + jX} = \frac{VR}{R^2 + X^2} - j\frac{VX}{R^2 + X^2}$$

Series circuit resonance occurs when $X_L = X_C$, at a frequency of

$$\omega_0 = 1/(LC)^{1/2} \qquad \text{rad/s}$$

or $\qquad f_0 = 1/2\pi(LC)^{1/2} \qquad \text{Hz}$

Impedance at resonance $= R$, and the circuit phase angle is zero.

Q-**factor of a series circuit**

$$Q = \text{voltage magnification at resonance}$$

$$= \frac{\omega_0 L}{R} = \frac{1}{\omega_0 CR} = \frac{1}{R}\left(\frac{L}{C}\right)^{1/2}$$

Conductance $\quad G = 1/R$ $\qquad\qquad\qquad$ S

Susceptance $\quad B = 1/X$ $\qquad\qquad\qquad$ S

Admittance $\quad Y = G + jB = |Y|\angle\phi$

where $\quad |Y| = (G^2 + B^2)^{1/2}$ and $\quad \phi = \tan^{-1} \dfrac{B}{G}$

Current in a parallel circuit $\quad I = VY = VG + jVB$

Parallel resonance This occurs when the reactive components of the current in both branches of the circuit are equal to one another. This resonant frequency is

$$\omega_0 = \left(\frac{1}{LC} - \frac{R^2}{4L^2} \right)^{1/2}$$

and if $\quad R \ll 2L,$ then $\quad \omega_0 = 1/(LC)^{1/2}$

The dynamic impedance of a parallel circuit at resonance is $R_D = L/CR \ \Omega$, and its phase angle is zero.

Q-factor of a parallel tuned circuit $\quad Q = \dfrac{\omega_0 L}{R} = \dfrac{R_D}{\omega_0 L}$

Impedance of a tuned circuit near resonance $Z = \dfrac{R_D}{1 + 2Q\delta}$

where $\quad \delta = (\omega \sim \omega_0)/\omega_0$

Volt-amperes (VA) $\quad S = EI \qquad$ (apparent power)

33

Power $P = EI \cos \phi = S \cos \phi$ (active power)

Reactive volt-amperes (VAr)

$$Q = EI \sin \phi = S \sin \phi \quad \text{(reactive power)}$$

Complex power $S = EI^*$

where I^* is the complex conjugate of the current (see the section on complex numbers). If

$E = E \underline{/\alpha}, I = I \underline{/\beta}$ and $\phi = (\alpha - \beta)$ then

$S = EI^* = E \underline{/\alpha} \times I \underline{/-\beta} = EI \underline{/(\alpha - \beta)} = EI\underline{/\phi}$

$\quad = EI \cos \phi + j \, EI \sin \phi$

$\quad = \text{power} + j(\text{reactive volt-amperes})$

Positive reactive volt-amperes represent lagging VAr, and negative reactive volt-amperes represent leading VAr.

Power factor $\cos \phi = \dfrac{P}{EI} = \dfrac{\text{kW}}{\text{kVA}}$

Current   active component $= I \cos \phi$

reactive component $= I \sin \phi$

Complex waves

$$e = E_{1m} \sin(\omega t + \phi_1) + E_{2m} \sin(2\omega t + \phi_2) + \cdots$$

R.M.S. value of a complex wave

$$E = \frac{1}{\sqrt{2}} \times (E_{1m}^2 + E_{2m}^2 + \cdots)^{1/2} = (E_1^2 + E_2^2 + \cdots)^{1/2}$$

where E_1, E_2, etc are the r.m.s. values of the harmonics.
Power supplied by a complex wave

$$P = E_1 I_1 \cos \phi_1 + E_2 I_2 \cos \phi_2 + \cdots$$

Reactance to complex waves

Inductance: reactance $= \omega L \times \left(\dfrac{E_1^2 + E_2^2 + E_3^2 + \cdots}{E_1^2 + \dfrac{E_2^2}{4} + \dfrac{E_3^2}{9} + \cdots} \right)^{1/2}$

Capacitance: reactance $= \dfrac{1}{\omega C} \times \left(\dfrac{E_1^2 + E_2^2 + E_3^2 + \cdots}{E_1^2 + 4E_2^2 + 9E_3^2 + \cdots} \right)^{1/2}$

Power factor with a complex wave

Power factor $= \dfrac{\text{Total power supplied}}{\text{Total r.m.s. voltage} \times \text{Total r.m.s. current}}$

$= \dfrac{V_1 I_1 \cos \phi_2 + V_2 I_2 \cos \phi_2 + \cdots + V_n I_n \cos \phi_n}{VI}$

THREE PHASE A.C. CIRCUITS

a = complex operator $1 \underline{/120°}$
E_L and I_L = line values
E_p and I_p = phase values
P = power
ϕ = phase angle

Voltage relationships

Three phase star $\quad E_L = \sqrt{3}\, E_p$
$\qquad\qquad\qquad\quad I_L = I_p$

Three phase delta $\quad E_L = E_p$
$\qquad\qquad\qquad\quad I_L = \sqrt{3}\, I_p$

Power

Power per phase = $V_p I_p \cos \phi$
Total power (balanced or unbalanced) = sum of the power
$\qquad\qquad\qquad\qquad\qquad$ consumed by each of the phases
Power consumed by a balanced load,

$$P = \sqrt{3}\, E_L I_L \cos \phi$$

VA consumed by a balanced load, $S = \sqrt{3}\, E_L I_L$
VAr consumed by a balanced load, $Q = \sqrt{3}\, E_L I_L \sin \phi$

Two wattmeter method of measuring power

under all conditions, $P = P_1 + P_2$

for a balanced load, $\cos \phi = \dfrac{1}{\left\{ 1 + 3 \left(\dfrac{P_1 - P_2}{P_1 + P_2} \right)^2 \right\}^{1/2}}$

$$\tan \phi = \sqrt{3} \left(\frac{P_1 - P_2}{P_1 + P_2} \right)$$

Operator a (or h) $a = 1 \underline{/120°}$

$$a^2 = 1 \underline{/240°}$$

$$a^3 = 1 \underline{/0°}$$

$$a + a^2 + a^3 = 0$$

$$a + a^2 + 1 = 0$$

Symmetrical components   In the following the subscript 0 indicates the zero sequence component, the subscript 1 indicates the positive phase sequence component, and the subscript 2 indicates the negative phase sequence component (see Figure 10).

$$I_R = I_{R0} + I_{R1} + I_{R2}$$

$$I_Y = I_{Y0} + I_{Y1} + I_{Y2} = I_{R0} + a^2 I_{R1} + a I_{R2}$$

$$I_B = I_{B0} + I_{B1} + I_{B2} = I_{R0} + a I_{R1} + a^2 I_{R2}$$

$$I_{R0} = (I_R + I_Y + I_B)/3$$

$$I_{R1} = (I_R + aI_Y + a^2I_B)/3$$

$$I_{R2} = (I_R + a^2I_Y + aI_B)/3$$

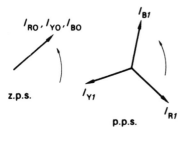
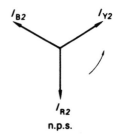

Figure 10

TRANSIENTS

C = capacitance
e = 2.71828 . . .
E = steady voltage
i = instantaneous value of current
L = inductance
q = instantaneous value of charge
R = resistance
t = time
v = voltage
v_0 = initial voltage across a capacitor
τ = time constant
ω_n = undamped natural frequency
ω_o = oscillatory frequency
ζ = damping factor

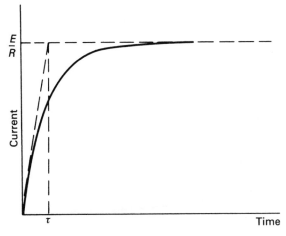

Figure 11

First-order circuits, step function response Inductive circuit (see Figure 11) with zero initial current

$$\text{voltage} = v = iR + L\frac{\mathrm{d}i}{\mathrm{d}t}$$

$$\text{current} = i = \frac{E}{R}\left(1 - \mathrm{e}^{-t/\tau}\right)$$

where $\tau = L/R$

$$\text{voltage across } R = v_R = E(1 - \mathrm{e}^{-t/\tau})$$

$$\text{voltage across } L = v_L = E\,\mathrm{e}^{-t/\tau}$$

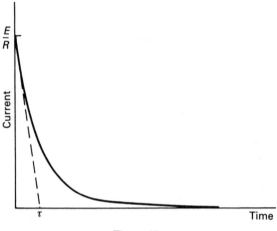

Figure 12

Capacitive circuit (see Figure 12) voltage $= v = iR + \dfrac{q}{C}$

$$\text{current} = i = \frac{E}{R}\,e^{-t/\tau}$$

where $\tau = RC$

$$\text{voltage across } R = v_R = E\,e^{-t/\tau}$$
$$\text{voltage across } C = v_C = E(1 - e^{-t/\tau})$$

Rise time (or fall time) between 10% and 90% for a step change in signal $= 2.2\tau$.

Second-order circuits, step function response R–L–C series circuit

$$\text{voltage} = v = L\frac{\mathrm{d}i}{\mathrm{d}t} + Ri + \frac{q}{C}$$

$$= L\left(\frac{\mathrm{d}i}{\mathrm{d}t} + \frac{R}{L}i + \frac{i\mathrm{d}t}{LC}\right)$$

if $\quad \omega_n = 1/(LC)^{\frac{1}{2}}$

$$\zeta = \text{damping factor} = \frac{\text{actual circuit resistance } R}{\text{resistance } R_d \text{ for critical damping}}$$

and $\quad R_d = 2(L/C)^{\frac{1}{2}}$

Zero damping ($\zeta = 0$ or $R_d = 0$)

$$i = \frac{E - v_o}{\omega_o L} \sin \omega_o t$$

Underdamped ($\zeta < 1$ or $R < R_d$)

$$i = \frac{E - v_o}{\omega_o L}\ e^{-\zeta \omega_n t} \sin \omega_o t$$

where $\quad \omega_o = \omega_n \sqrt{(1 - \zeta^2)}$

Critical damping ($\zeta = 1$ or $R = R_d$)

$$i = \frac{E - v_o}{L}\ t\, e^{-\omega_n t}$$

41

Overdamped ($\zeta > 1$ or $R > R_d$)

$$i = \frac{E - v_o}{2L\omega_o} e^{-\zeta\omega_n t}(e^{\omega_o t} - e^{-\omega_o t})$$

where $\omega_o = \omega_n \sqrt{(\zeta^2 - 1)}$

RECTIFICATION

E = supply r.m.s. voltage
E_A = forward p.d. across rectifier
E_L = average load voltage
E_m = maximum supply voltage
f = supply frequency
α = delay angle of conduction
β = extinction angle
ω = angular frequency of supply

Single-phase half-wave controlled   (see Figure 13)

$$E_L = \frac{1}{2\pi}\int_\alpha^\beta (E_m \sin \omega t - E_A)\, d(\omega t)$$

If $E_A = 0$ and $\beta = \pi$, $E_L = \dfrac{E_m}{2\pi}(1 + \cos \alpha)$

$$= 0.225E(1 + \cos \alpha)$$

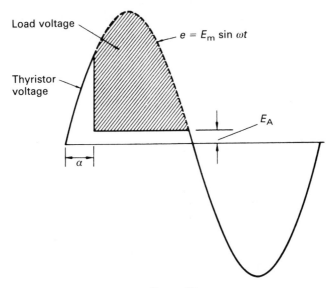

Figure 13

Ripple

$$\text{per-unit ripple} = \frac{\text{r.m.s. fundamental ripple voltage}}{E_L}$$

Fourier expressions

half wave $E_m\left(\dfrac{1}{\pi} + \dfrac{1}{2}\sin \omega t - \dfrac{2}{\pi} \displaystyle\sum_{n=2,\,4,\,6,\,\ldots} \dfrac{\cos n\omega t}{(n+1)(n-1)}\right)$

full wave $E_m\left(\dfrac{2}{\pi} - \dfrac{4}{\pi} \displaystyle\sum_{n=2,\,4,\,6,\,\ldots} \dfrac{\cos n\omega t}{(n+1)(n-1)}\right)$

Simple R—C ripple filter   ripple voltage $= \delta V_L = E_m T / R_L C$
where R_L = load resistance and C = capacitance

$$\text{load voltage} = V_L = E_m - \delta V_L / 2$$

Rectifier circuit data

Single phase

    half wave $E_L = 0.318\ E_m = 0.45\ E$
                 fundamental ripple frequency $= f$
                 per-unit ripple $= 1.11$

    full wave centre-tap
                 $E_L = 0.636\ E_m = 0.9\ E$
                 fundamental ripple frequency $= 2f$
                 per-unit ripple $= 0.472$

    full wave bridge
                 $E_L = 0.636\ E_m = 0.9\ E$
                 fundamental ripple frequency $= 2f$
                 per-unit ripple $= 0.472$

*Three phase**

    half wave $E_L = 1.17\ V_p$
                 fundamental ripple frequency $= 3f$
                 per-unit ripple $= 0.177$

    full wave centre-tap
                 $E_L = 1.35\ V_p$
                 fundamental ripple frequency $= 6f$
                 per-unit ripple $= 0.04$

full wave bridge
$E_L = 1.35\ V_L = 2.34\ V_p$

fundamental ripple frequency $= 6f$

per-unit ripple $= 0.04$

double-star $E_L = 1.17\ V_p$

fundamental ripple frequency $= 6f$

per-unit ripple $= 0.04$

$^*V_p$ = phase voltage

V_L = line voltage

BRIDGE MEASURING CIRCUITS

Name of circuit	Circuit	Parameter measured	Value of unknown constants
Wheatstone	Figure 14	Resistance	$Z = S\dfrac{Q}{P}$
De Sauty	Figure 15	Capacitors with negligible loss	$C = C_2\dfrac{P}{Q}$
Schering	Figure 16	Capacitance and loss angle	$C = C_3\dfrac{R_2}{R_1};\ R = \dfrac{R_1 C_2}{C_3}$ $\delta = \tan^{-1}\omega C_2 R_2$
Owen	Figure 17	Inductance with resistance	$L = C_1 R_2 R_3;$ $R = \dfrac{R_2 C_1}{C_3}$
Wien	Figure 18	Frequency	$\omega^2 = \dfrac{1}{R_3 R_4 C_3 C_4}$
Transformer ratio-arm bridge	Figure 19	Admittance	$G_2 = G_1 N_1/N_2$ $B_2 = B_1 N_1/N_2$

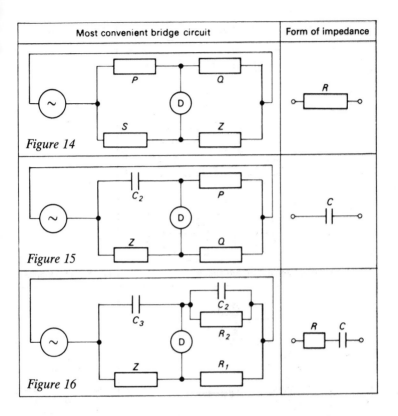

Most convenient bridge circuit	Form of impedance
Figure 14	R
Figure 15	C
Figure 16	R C

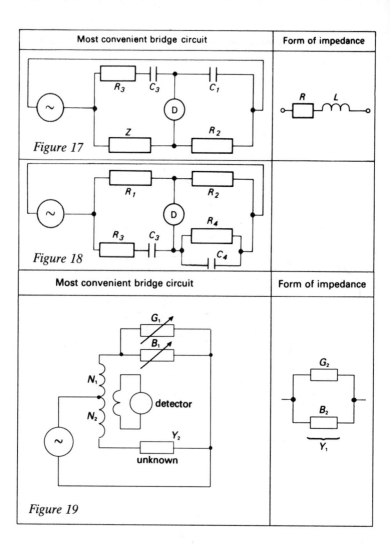

Most convenient bridge circuit	Form of impedance
Figure 17	R L
Figure 18	

Most convenient bridge circuit	Form of impedance
Figure 19	G_2 B_2 Y_1

INSTRUMENTS

B = flux density
d = mean diameter of coil
f = supply frequency
F = viscous friction coefficient
G = constant of the movement
I = current
J = inertia
K = spring constant
l = active length of conductor
N = number of turns
Q = quantity of electricity
R = resistance of the coil and the circuit
R_f = resistance of the former
T = torque
V = voltage
θ = angular deflection
θ_1 = initial deflection
Φ = magnetic flux
ω_n = undamped natural frequency
ω_o = oscillatory frequency
ζ = damping factor

Moving-coil instruments   Equation of motion

$$J \frac{d^2\theta}{dt^2} + F \frac{d\theta}{dt} + K\theta = GI \qquad \text{Nm}$$

where $G = NBld$

Steady-state deflection $\theta_{ss} = \dfrac{G}{K} I$

49

Torque acting on N turns $T = GI$

Total electromagnetic damping

$$\text{damping} = G^2 \left(\frac{1}{R} + \frac{1}{N^2 R_f} \right) \text{Nm/(rad/s)}$$

Peak overshoot for an underdamped movement
after a time π/ω_n, per-unit overshoot $= e^{-\zeta\pi/(1-\zeta^2)^{1/2}}$
where $\omega_n = (K/J)^{1/2}$ and $\zeta = F/2(KJ)^{1/2}$

Ballistic galvanometer   solution of equation of motion

$$\theta = \frac{GQ}{J\omega_o} e^{-\zeta\omega_n t} \sin \omega_n t$$

Charge measured $Q = \dfrac{K}{G} \dfrac{\theta_1}{\omega_n}$

Fluxmeter $\Delta\Phi = \dfrac{K}{N} \Delta\theta$

Induction instruments   ammeters and voltmeters $T \propto I^2 f$
energy meters $T \propto VI \cos \phi$

LOGIC

Basic theorems

$A.1 = A$	$A + 1 = 1$
$A.0 = 0$	$A + 0 = A$
$A.A = A$	$A + A = A$
$A.\bar{A} = 0$	$A + \bar{A} = 1$
$\bar{\bar{A}} = A$	

De Morgan's theorem

$$\overline{A + B} = \bar{A}.\bar{B}$$
$$\overline{A.B} = \bar{A} + \bar{B}$$

Laws   Commutative law $A + B = B + A$ $A.B = B.A$
Associative law $A + (B + C) = (A + B) + C$
$$A.(B.C) = (A.B).C$$
Distributive law $A.(B + C) = A.B + A.C$
$$A + (B.C) = (A + B).(A + C)$$

LOGIC GATES

OR, AND, NOT, NOR, NAND functions   The logic OR connective is represented either by a '+' or a '$\cup$'.

The logic AND connective is represented either by a '.' or a '$\cap$'.

If the output from a logic gate is represented by f, then for

OR gate	$f = A + B + \ldots + M + N$
AND gate	$f = A.B. \ldots .M.N$
NOT gate	$f = \overline{A}$
NOR gate	$f = \overline{A + B + \ldots + M + N}$
NAND gate	$f = \overline{A.B. \ldots .M.N}$

Truth tables for two-input gates

Inputs	Output			
A B	OR	NOR	AND	NAND
0 0	0	1	0	1
0 1	1	0	0	1
1 0	1	0	0	1
1 1	1	0	1	0

VALUES OF $2^{\pm N}$

2^N	N	2^{-N}
1	0	1
2	1	0.5
4	2	0.25
8	3	0.125
16	4	0.062 5
32	5	0.031 25
64	6	0.015 625
128	7	0.007 812 5
256	8	0.003 906 25
512	9	0.001 953 125
1024	10	0.000 976 562 5
2048	11	0.000 488 281 25
4096	12	0.000 244 140 625

EQUIVALENT CIRCUITS AND FOUR-TERMINAL NETWORK EQUATIONS

h-parameters (see Figure 20) $\quad V_1 = h_i I_1 + h_r V_2$
$$I_2 = h_f I_1 + h_o V_2$$

h_i is the input parameter, having dimensions of resistance
h_r is the reverse feedback parameter, and is dimensionless
h_f is the forward current gain, and is dimensionless
h_o is the output parameter, having dimensions of conductance

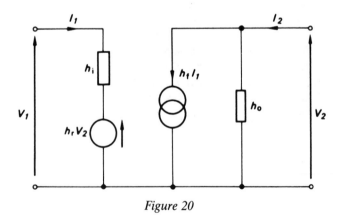

Figure 20

$$h_i = \left(\frac{\delta V_1}{\delta I_1}\right)_{\delta V_2=0} \qquad\qquad h_r = \left(\frac{\delta V_1}{\delta V_2}\right)_{\delta I_1=0}$$

$$h_f = \left(\frac{\delta I_2}{\delta I_1}\right)_{\delta V_2=0} \qquad\qquad h_o = \left(\frac{\delta I_2}{\delta V_2}\right)_{\delta I_1=0}$$

y-parameters (see Figure 21) $\quad I_1 = y_i V_1 + y_r V_2$
$$I_2 = y_f V_1 + y_o V_2$$

All the y-parameters have dimensions of admittance.

$$y_i = \left(\frac{\delta I_1}{\delta V_1}\right)_{\delta V_2=0} \qquad\qquad y_r = \left(\frac{\delta I_1}{\delta V_2}\right)_{\delta V_1=0}$$

$$y_f = \left(\frac{\delta I_2}{\delta V_1}\right)_{\delta V_2=0} \qquad\qquad y_o = \left(\frac{\delta I_2}{\delta V_2}\right)_{\delta V_1=0}$$

N.B. The symbols g_f and g_m are sometimes used to replace y_f, and the symbol y_o is sometimes replaced by $1/r_d$.

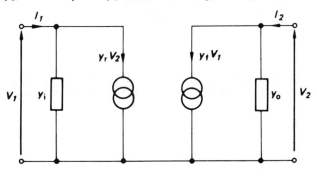

Figure 21

55

ABCD parameters
$$V_1 = AV_2 + BI_2$$
$$I_1 = CV_2 + DI_2$$

A is dimensionless
B has the dimensions of impedance
C has the dimensions of admittance
D is dimensionless

$$A = \left(\frac{\delta V_1}{\delta V_2}\right)_{\delta I_2 = 0} \qquad B = \left(\frac{\delta V_1}{\delta I_2}\right)_{\delta V_2 = 0}$$

$$C = \left(\frac{\delta I_1}{\delta V_2}\right)_{\delta I_2 = 0} \qquad D = \left(\frac{\delta I_1}{\delta I_2}\right)_{\delta V_2 = 0}$$

Bilateral networks   Elements such as resistors, inductors and capacitors are *bilateral elements* that may be connected in either direction and the circuit is unchanged. Networks containing only bilateral elements are known as *bilateral networks* or *reciprocal networks* and, for these networks, the parameter $y_r = y_f$ and the parameter $B = C$. The following relationship holds good.

$$AD - BC = 1$$

ELECTRONIC AMPLIFIERS

K_i = current gain
K_p = power gain
K_v = voltage gain
R_D = dynamic resistance of a tuned circuit
R_{in} = input resistance
R_L = load resistance

h_i, h_r, h_f, h_o = general h-parameters
h_{ie}, h_{re}, h_{fe}, h_{oe} = common-emmitter connection hybrid parameters
y_i, y_r, y_f, y_o = general y-parameters
y_{is}, y_{rs}, y_{fs}, y_{os} = common-source connection y-parameters
R_{out} = output resistance
R_S = internal resistance of source
Y_L = load admittance
Y_S = internal admittance of source

Relationships between the general h and y-parameters

$$h_i = 1/y_i \qquad\qquad y_i = 1/h_i$$

$$h_r = -y_r/y_i \qquad\qquad y_r = -h_r/h_i$$

$$h_f = y_f/y_i \qquad\qquad y_f = h_f/h_i$$

$$h_o = y_o - \frac{y_r y_f}{y_i} \qquad y_o = h_o - \frac{h_r h_f}{h_i}$$

Relationships between the common-emitter, common-base and common-collector h-parameters

$$h_{ib} = \frac{h_{ie}}{1 + h_{fe}} \qquad h_{ic} = h_{ie}$$

$$h_{rb} = \frac{h_{ie} h_{oe}}{1 + h_{fe}} \qquad h_{rc} = \frac{1}{1 + h_{re}}$$

$$h_{fb} = \frac{-h_{fe}}{1 + h_{fe}} \qquad h_{fc} = -(1 + h_{fe})$$

$$h_{ob} = \frac{h_{oe}}{1 + h_{fe}} \qquad h_{oc} = h_{oe}$$

57

Relationships between y, h and $ABCD$ parameters

	y		h		$ABCD$	
y	y_i	y_r	$\dfrac{1}{h_i}$	$\dfrac{-h_r}{h_i}$	$\dfrac{D}{B}$	$\dfrac{-D_{ABCD}}{B}$
	y_f	y_o	$\dfrac{h_f}{h_i}$	$\dfrac{D_h}{h_i}$	$\dfrac{-1}{B}$	$\dfrac{A}{B}$
h	$\dfrac{1}{y_i}$	$\dfrac{-y_r}{y_i}$	h_i	h_r	$\dfrac{B}{D}$	$\dfrac{D_{ABCD}}{D}$
	$\dfrac{y_f}{y_i}$	$\dfrac{D_y}{y_i}$	h_f	h_o	$\dfrac{-1}{D}$	$\dfrac{C}{D}$
$ABCD$	$\dfrac{-y_o}{y_f}$	$\dfrac{-1}{y_f}$	$\dfrac{D_h}{h_f}$	$\dfrac{h_i}{h_f}$	A	B
	$\dfrac{-D_y}{y_f}$	$\dfrac{-y_i}{y_f}$	$\dfrac{-h_o}{h_f}$	$\dfrac{-1}{h_f}$	C	D

For all parameter sets

$$D_p = P_i P_o - P_r P_f$$

For example, $\quad D_h = h_i h_o - h_r h_f$

58

Common-emitter amplifier equations

Quantity	Exact relationship	Approximate relationship
K_i	$\dfrac{h_{fe}}{1 + h_{oe}R_L}$	h_{fe}
R_{in}	$h_{ie} - h_{re}R_L K_i$	h_{ie}
K_v	$-K_i R_L/R_{in}$	$-h_{fe}R_L/h_{ie}$
R_{out}	$1/(h_{oe} - \dfrac{h_{fe}h_{re}}{h_{ie} + R_S})$	$1/h_{oe}$
K_p	$K_i^2 R_L/R_{in} = |K_i K_v|$	$h_{fe}^2 R_L/h_{ie} = |K_i K_v|$

Common-base amplifier equations

$$K_i = h_{fb}/(1 + h_{ob}R_L)$$

$$R_{in} = h_{ib} - h_{rb}R_L K_i$$

$$K_v = -K_i R_L/R_{in}$$

$$R_{out} = 1/\left(h_{ob} - \frac{h_{fb}h_{rb}}{h_{ib} + R_S}\right)$$

$$K_p = K_i^2 R_L/R_{in} = |K_i K_v|$$

Common-collector amplifier equations (emitter follower)

$$K_i = h_{fc}/(1 + h_{oc}R_L) \approx -(1 + h_{fe})$$

$$R_{in} = h_{ic} - h_{rc}R_L K_i \approx h_{ie} + (1 + h_{fe})R_L$$

$$K_v = -K_i R_L/R_{in} \approx 1$$

$$R_{\text{out}} = 1/ \left(h_{\text{oc}} - \frac{h_{\text{fc}} h_{\text{rc}}}{h_{\text{ic}} + R_{\text{S}}} \right) \approx \frac{h_{\text{ie}} + R_{\text{S}}}{1 + h_{\text{fe}}}$$

$$K_{\text{p}} = K_{\text{i}}^2 R_{\text{L}}/R_{\text{in}} \approx 1 + h_{\text{fe}}$$

Common-source amplifier equations

Quantity	Exact relationship	Approximate relationship
K_{v}	$-\dfrac{y_{\text{fs}}}{y_{\text{os}} + Y_{\text{L}}}$	$-y_{\text{fs}}/Y_{\text{L}}$
K_{i}	$\dfrac{y_{\text{fs}} + K_{\text{v}} y_{\text{os}}}{y_{\text{is}} + K_{\text{v}} y_{\text{rs}}}$	infinity
K_{p}	$\|K_{\text{v}} K_{\text{i}}\|$	infinity
R_{in}	$1/(y_{\text{is}} + K_{\text{v}} y_{\text{rs}})$	$1/y_{\text{is}}$
R_{out}	$1/ \left(y_{\text{os}} - \dfrac{y_{\text{fs}} y_{\text{rs}}}{Y_{\text{S}} + y_{\text{is}}} \right)$	$1/y_{\text{os}}$

N.B. The symbol y_{fs} is sometimes replaced by g_{fs} or g_{m}, and y_{os} is sometimes replaced by $1/r_{\text{d}}$.

Tuned collector amplifier:

$$\text{voltage gain at resonance} = \frac{-h_{\text{fe}} R_{\text{D}}}{h_{\text{ie}}(1 + h_{\text{oe}} R_{\text{D}})}$$

FEEDBACK AMPLIFIERS

A = amplifier gain without feedback
A' = amplifier gain with feedback
e_{out}, e_1 = instantaneous values of e.m.f.
R_1, R_2, R_f = values of resistance
β = feedback factor

Effects of feedback The primary effect of the type of feedback used, that is, positive or negative, is a change in gain.

Type of feedback		General effect on gain	General effect on R_{out}	General effect on R_{in}
Positive Negative		increased reduced		
Voltage (shunt derived)	positive		increased	
	negative		reduced	
Current (series derived)	positive		reduced	
	negative		increased	
Shunt injected	positive			increased
	negative			reduced
Series injected	positive			reduced
	negative			increased

The primary effect of the way in which feedback is derived, that is, voltage or current, is a change in output resistance.

The primary effect of the way in which the feedback is applied, that is, series or shunt, is a change in input resistance.

Voltage gain of a series voltage feedback amplifier (see Figure 22)

$$A' = \frac{A}{1 - A\beta}$$

For a negative feedback amplifier, if $A\beta \gg 1$, then $A' = -1/\beta$.

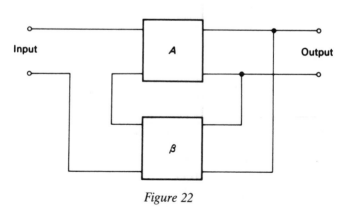

Figure 22

Summing amplifier (see Figure 23)   If the amplifier gain
A is large, then

$$V_{\text{out}} = -R_{\text{f}} \left(\frac{V_1}{R_1} + \frac{V_2}{R_2} + \ldots \right)$$

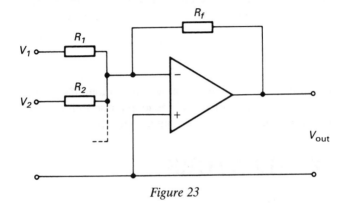

Figure 23

Figure 24

63

Integrator (see Figure 24) $e_{\text{out}} = -\dfrac{1}{CR} \displaystyle\int_0^t V_1 \, dt$

Effect of feedback on bandwidth   if B_1 and B_2 are the values of bandwidth without and with feedback, respectively, and G_1 and G_2 are the respective gains, then

$$G_1 B_1 = G_2 B_2$$

Conditions for instability in a closed-loop amplifier   the loop gain has unity value and the loop phase shift is zero, the two conditions occurring simultaneously.

OSCILLATORS

C = capacitance
L = inductance
M = mutual inductance
R = resistance
ω_o = oscillatory frequency

Conditions for maintenance of oscillations   the loop gain must be unity and the loop phase shift must be zero, both conditions being satisfied simultaneously.

Oscillators

Colpitts oscillator (Figure 25)

$$\omega_o = \left\{ \frac{1}{L}\left(\frac{1}{C_1} + \frac{1}{C_2} \right) \right\}^{1/2}$$

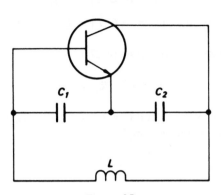

Figure 25

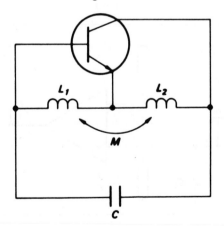

Figure 26

Hartley oscillator (Figure 26)

$$\text{if } M = 0, \ \omega_o = 1/\{C(L_1 + L_2)\}^{1/2}$$
$$\text{if } M \neq 0, \ \omega_o = 1/\{C(L_1 + L_2 + 2M)\}^{1/2}$$

Wien bridge oscillator (Figure 27)

$$\omega_o = 1/(R_1 R_2 C_1 C_2)^{1/2}$$

and if $R = R_1 = R_2$ and $C = C_1 = C_2$, then $\omega_o = 1/RC$
 For oscillation the amplifier gain = 3

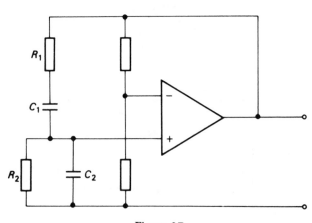

Figure 27

R–C ladder network oscillator

$$\omega_o = 1/(CR\sqrt{6})$$

and for maintenance of oscillations, $h_{fe} \geqslant 29$

Symmetrical astable multivibrator   pulse repetition frequency $= 1/(1.38\ CR)$

Monostable multivibrator   pulse width $= 0.69CR$

CONTROL SYSTEMS

F = viscous friction coefficient
J = inertia
K = system gain
T_L = external load torque
ϵ = error angle
ω = shaft speed in rad/s
ω_o = oscillatory frequency
ω_n = undamped natural frequency
θ_i, θ_o = angular displacement
ζ = damping factor

Equation of motion of a second-order r.p.c. servosystem

$$J\frac{d^2\theta_o}{dt^2} + F\frac{d\theta_o}{dt} + K\theta_o = K\theta_i$$

Standard form of equation of motion

$$\frac{d^2\theta_o}{dt^2} + 2\zeta\omega_n\frac{d\theta_o}{dt} + \omega_n^2\theta_o = \omega_n^2\theta_i$$

where

$$\omega_n = (K/J)^{1/2}$$

and

$$\zeta = F/2(KJ)^{1/2}$$

Response to a step input

Zero damping ($\zeta = 0$)

$$\theta_o = \theta_i(1 - \cos \omega_n t)$$

Underdamped ($\zeta < 1$)

$$\theta_o = \theta_i \left\{ 1 - e^{-\zeta\omega_n t}\left(\cos \omega_o t + \frac{\zeta}{(1 - \zeta^2)^{1/2}} \sin \omega_o t \right) \right\}$$

where

$$\omega_o = \omega_n(1 - \zeta^2)^{1/2}$$

Maximum overshoot occurs at $t = \pi/\omega_o$, and is $e^{-\zeta\pi/(1 - \zeta^2)^{1/2}}$ per unit

Critical damping ($\zeta = 1$)

$$\theta_{o} = \theta_{i} \{1 - e^{-\omega_n t}(1 + \omega_n t)\}$$

Overdamped ($\zeta > 1$)

$$\theta_{o} = \theta_{i} \left\{ 1 - e^{-\zeta \omega_n t}\left(\cosh \beta t + \frac{\zeta}{(\zeta^2 - 1)^{\frac{1}{2}}} \sinh \beta t \right) \right\}$$

where

$$\beta = \omega_n (\zeta^2 - 1)^{\frac{1}{2}}$$

Response to a ramp input ($\theta_i = \omega t$)

Underdamped

$$\theta_{o} = \omega t - \frac{2\zeta\omega}{\omega_n} + \frac{2\zeta\omega}{\omega_n} e^{-\zeta\omega_n t}\left(\cos \omega_o t + \frac{2\zeta^2 - 1}{2\zeta(1 - \zeta^2)^{\frac{1}{2}}} \sin \omega_o t \right)$$

Critical damping

$$\theta_{o} = \omega t - \frac{2\omega}{\omega_n} + \frac{2\omega}{\omega_n} e^{-\omega_n t}\left(1 + \frac{\omega_n t}{2} \right)$$

Overdamped

$$\theta_{o} = \omega t - \frac{2\zeta\omega}{\omega_n}\left\{ 1 - e^{-\zeta\omega_n t}\left(\cosh \beta t + \frac{2\zeta^2 - 1}{2\zeta(\zeta^2 - 1)^{\frac{1}{2}}} \sinh \beta t \right) \right\}$$

Velocity lag in an underdamped second-order system

$$\epsilon = \frac{F\omega_1}{K}\,\text{rad}$$

Effect of load torque on a simple system

$$\text{Offset} = \frac{T_L}{K}\,\text{rad}$$

Simplified Nyquist criterion for stability

The $(-1 + j0)$ point must lie on the left of the Nyquist locus as it is traversed in the direction of increasing frequency.

D.C. MACHINES

c = number of parallel paths between the brushes
E = e.m.f.
I_a = armature current
n = speed in revolutions/s
p = number of pole pairs
R_a = armature resistance
T = torque
Z = number of conductors
Φ = flux
η = per-unit efficiency
ω = speed in rad/s

Basic relationships

$$E \propto \Phi n \qquad T \propto \Phi I_a$$

$$E = \frac{2p}{c} \Phi Z n$$

$$E I_a = T \omega$$

where $\quad c = 2$ for wave windings

and $\qquad c = 2p$ for lap windings

$$T = \frac{1}{\pi} \frac{p \Phi Z I_a}{c} \qquad\qquad \text{Nm}$$

For a generator $\qquad V = E - I_a R_a$

For a motor $\qquad\quad V = E + I_a R_a$

Per-unit efficiency $\; = \dfrac{\text{output}}{\text{output} + \text{losses}}$

TRANSFORMERS

$\cos \phi$ = load power factor
E = r.m.s. induced e.m.f. per phase
f = frequency
I = current
N = number of turns
P_i = iron loss

71

R_1, R_2 = resistance of the primary and secondary
windings, respectively.
R_{e1} = equivalent resistance referred to the primary
R_{e2} = equivalent resistance referred to the secondary
V = r.m.s. voltage per phase
X_1, X_2 = reactances of the primary and secondary
windings
X_e = equivalent reactance
Φ = magnetic flux

The following relationships hold good in an *ideal transformer* or *power transformer*, in which the windings are on an iron core and the magnetic coupling coefficient is practically unity.

E.M.F. equation $\quad E = 4.44\, f\Phi N$

Turns ratio $\quad \dfrac{N_2}{N_1} \approx \dfrac{E_2}{E_1} \approx \dfrac{I_1}{I_2}$

Equivalent resistance referred to the primary

$$R_{e1} = R_1 + R_2 \left(\frac{N_1}{N_2}\right)^2$$

Equivalent leakage reactance referred to the primary

$$X_{e1} = X_1 + X_2 \left(\frac{N_1}{N_2}\right)^2$$

Voltage regulation

$$\text{per-unit regulation} = \frac{\text{no-load voltage} - \text{full-load voltage}}{\text{no-load voltage}}$$

Efficiency

$$\text{per-unit efficiency} = \frac{V_2 I_2 \cos \phi}{V_2 I_2 \cos \phi + I_2^2 R_{e2} + P_i}$$

maximum efficiency occurs when copper loss = iron loss

The linear transformer   For a magnetically coupled circuit with mutual inductance M between the windings

self-impedance of the primary $\quad = Z_{11} = R_1 + j\omega L_1$

self-impedance of the secondary $\; = Z_{22}$

$$= R_2 + j\omega L_2 + Z_L$$

where Z_L is the impedance of the load, and

$$\text{input impedance} = Z_{in} = Z_{11} + \frac{\omega^2 M^2}{Z_{22}}$$

where ω is the supply frequency.

SYNCHRONOUS MACHINES

V = output voltage
E_0 = no-load output voltage
f = frequency
g = slots per pole per phase
I_p = phase current
k_d = distribution factor
k_s = coil-span factor
n = speed in revolutions/s
N_p = number of turns in series per phase
p = number of pole pairs
P = power
T = torque
Z = number of conductors in series per phase
ψ = slot phase difference
σ = 180° – angle of coil span
Φ = flux per pole
ω = electrical speed of the shaft = $p\omega_r$
ω_r = speed of rotation
ϕ = phase angle between the induced e.m.f. and the winding current

Frequency $f = np$

Distribution factor or breadth factor

$$k_d = \frac{\text{e.m.f. with a distributed winding}}{\text{e.m.f. with a concentrated winding}} = \frac{\sin g\dfrac{\psi}{2}}{g\sin\dfrac{\psi}{2}}$$

Coil-span factor or pitch factor

$$k_s = \frac{\text{e.m.f. with a short-pitch coil}}{\text{e.m.f. with a full-pitch coil}} = \cos\frac{\sigma}{2}$$

E.M.F. per phase $E_0 = 4.44\, k_d k_s f \Phi N_p$

$$= k_d k_s \omega \Phi N_p / \sqrt{2}$$

$$= 2.22\, k_d k_s f \Phi Z$$

Voltage regulation per-unit regulation $= (E_0 - V)/V$

Torque per phase $T = \dfrac{p}{\sqrt{2}}\, N_p \Phi_m k_d k_s I_p \cos\phi$ N m

Mechanical power $P = \omega_r T$

INDUCTION MACHINES

E = rotor e.m.f.
E_0 = standstill rotor e.m.f.
f = supply frequency
f_r = frequency of rotor current
m = number of phases
n_r = speed of rotation in rev/s
n_s = synchronous speed in rev/s
p = number of pole pairs
P_I = input power

P_m = mechanical power
P_r = rotor copper loss
R = rotor resistance per phase
s = per-unit slip
T = gross torque
X_0 = standstill reactance per phase
Z_r = rotor impedance per phase
ω_s = synchronous speed in rad/s
Φ = flux per pole

Speed of rotation of the magnetic field $n_s = \dfrac{f}{p}$

Per-unit slip $s = \dfrac{n_s - n_r}{n_s}$

Frequency of rotor current $f_r = sf$

Rotor e.m.f. per phase $E = sE_0$

Rotor impedance per phase $Z_r = \{R^2 + (sX_0)^2\}^{\frac{1}{2}}$

Input power $P_I = \omega_s T$

Rotor copper loss $P_r = sP_I$

Rotor output power $P_m = \omega_r T$
 = rotor input power $\times\, n_r/n_s$
 = $P_I - P_r$

76

$$P_\mathrm{I}:P_\mathrm{r}:P_\mathrm{m} = 1:s:(1 - s)$$

Torque developed $\quad T = \dfrac{mE_0^{\ 2}}{\omega_\mathrm{s}} \ \dfrac{sR}{[R^2 + (sX_0)^2]}$

$$\propto \dfrac{\Phi^2 sR}{R^2 + (sX_0)^2}$$

Condition for maximum torque $\quad$ occurs when $R = sX_0$

Starting torque $\quad$ starting torque is proportional to (stator applied voltage)^2

Circle diagram (see Figure 28)

The following applies for a motor operating at point G.

To evaluate per phase	Multiply V_1 by
standstill rotor copper loss	JK
standstill stator copper loss	IJ
constant loss	HI or CD or AB
standstill input power	HK
stator input power	CG
stator copper loss	DE
stator ouput (rotor input) power	EG
rotor copper loss	EF
rotor output power	FG
rotor torque	$\begin{cases} FG/2\pi n_\mathrm{r} \\ EG/2\pi n_\mathrm{s} \end{cases}$

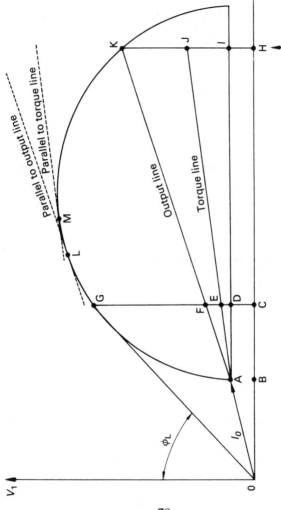

Figure 28

78

$$n_r = n_s \times \frac{FG}{EG}$$

$$s = \frac{EF}{EG}$$

$$\text{per-unit efficiency} = \frac{FG}{CG}$$

$$cos\ \phi_L = \frac{CG}{OG}$$

where O is the origin of the diagram.
Maximum power output occurs at point L
Maximum torque is developed at point M

THE PER-UNIT SYSTEM

For any quantity A $A_{pu} = \dfrac{A}{A_{base}}$

Rating

$$S_{base} = V_{base} \times I_{base} \qquad \text{VA or W}$$

Current

$$I_{base} = V_{base}\, Y_{base} = V_{base}\, G_{base} = V_{base}\, B_{base}$$

Voltage

$$V_{base} = I_{base} Z_{base} = I_{base} R_{base} = I_{base} X_{base}$$

Change of base $A_{pu1} \times A_{base1} = A_{pu2} \times A_{base2}$

POWER SYSTEM SHORT CIRCUITS

Base values The rated full-load voltage, current, and volt-amperes are chosen as the base values.

Reactance

$$X_{pu} = \frac{X}{X_{base}} = X \times \frac{I_{base}}{V_{base}}$$

Short-circuit current from a generator of reactance X

$$I_{sc} = \frac{V_{base}}{X} = \frac{I_{base}}{X_{pu}}$$

Short-circuit VA

$$S_{sc} = \frac{S_{base}}{X_{pu}}$$

Resistance

$$R_{\text{pu}} = R \times \frac{I_{\text{base}}}{V_{\text{base}}}$$

Short-circuit VA in a network of resistance R

$$S_{\text{sc}} = \frac{S_{\text{base}}}{R_{\text{pu}}}$$

Short-circuit VA in a network containing series R and X

$$S_{\text{sc}} = \frac{S_{\text{pu}}}{Z_{\text{pu}}} = \frac{S_{\text{pu}}}{R_{\text{pu}} + jX_{\text{pu}}}$$

Change of base
Let Z_{puM} be the per-unit value of Z_{M} referred to base M, and $Z_{\text{puM}'}$ be the per-unit value of Z_{M} referred to base N.

$$Z_{\text{puM}} \, Z_{\text{baseM}} = Z_{\text{puM}'} \, Z_{\text{baseN}}$$

or

$$Z_{\text{puM}'} = Z_{\text{puM}} \, \frac{Z_{\text{baseM}}}{Z_{\text{baseN}}}$$

$$= Z_{\text{puM}} \, \frac{V_{\text{baseM}}}{I_{\text{baseM}}} \, \frac{I_{\text{baseN}}}{V_{\text{baseN}}}$$

and if $V_{\text{baseM}} = V_{\text{baseN}}$, then

$$Z_{\text{puM}'} = Z_{\text{puM}} \, \frac{S_{\text{baseN}}}{S_{\text{baseM}}}$$

81

MODULATION

v = instantaneous value of e.m.f.
V_m = maximum value of e.m.f.
m = modulation of factor or index
p = modulation frequency
ω = carrier frequency

Amplitude modulation $v = V_m \sin \omega t (1 + m \sin pt)$
where m = depth of modulation or modulation factor

$$\text{Transmission efficiency} = \frac{\text{sideband power}}{\text{total power}} = \frac{m^2}{2 + m^2}$$

Phase modulation $v = V_m \sin (\omega t + m_p \sin pt)$ where
m_p = modulation index, which is proportional to p.

Frequency modulation $v = V_m \sin (\omega t + m_f \sin pt)$ where
m_f = modulation index

$$= \frac{\text{frequency deviation}}{\text{modulating frequency}} = \text{constant} \times \frac{\delta \omega}{p}$$

and is proportional to $1/p$.

TRANSMISSION LINES

C, G, L, R = coefficients per unit length of line
V = r.m.s. voltage at point x
V_s = r.m.s. sending-end voltage
I = r.m.s. current at point x
I_s = r.m.s. sending-end current
γ = propagation coefficient
x = distance from sending end
Z_o = characteristic impedance
α = attenuation coefficient (nepers/unit length)
β = phase coefficient (radians/unit length)
λ = wavelength of line

Line equations
$$V = V_s \cosh \gamma x - (I_s Z_o) \sinh \gamma x$$
$$I = I_s \cosh \gamma x - (V_s/Z_o) \sinh \gamma x$$

where $\quad \gamma = \{(R + j\omega L)/(G + j\omega C)\}^{\frac{1}{2}} = \alpha + j\beta$

Exponential forms of the equations
$$V = A\, e^{\gamma x} + B\, e^{-\gamma x}$$
$$I = C\, e^{\gamma x} + D\, e^{-\gamma x}$$
$$= \frac{A}{Z_o} e^{\gamma x} - \frac{B}{Z_o}\, e^{-\gamma x}$$

where $\quad Z_o = \{(R + j\omega L)/(G + j\omega C)\}^{\frac{1}{2}}$

and $\quad A = D = \cosh \gamma l$
$$B = Z_o \sinh \gamma l$$
$$C = \frac{1}{Z_o} \cosh \gamma l$$

Phase velocity of transmission $\quad u = \omega/\beta$

Wavelength $\quad \lambda = 2\pi/\beta$

ALGEBRA

Mensuration

Triangle   Area $= \frac{1}{2} \times$ base $\times$ perpendicular height

$$= \frac{1}{2} bc \sin A = \frac{1}{2} ca \sin B = \frac{1}{2} ab \sin C$$

$$= \sqrt{(s(s - a)(s - b)(s - c))}$$

$$\text{where } s = \frac{1}{2} (a + b + c)$$

Circle   Area $= \pi r^2$

Circumference $= 2\pi r = \pi d$

Area of sector $= \frac{1}{2} r^2 \theta$

Length of arc $= r\theta$ (θ in radians)

Ellipse   Area $= \pi ab$

Perimeter $= \pi(a + b)$

Cylinder   Volume $= \pi r^2 h$

Area of curved surface $= 2\pi rh$

Pyramid   Volume $=$ area of base $\times$ height/3

Cone   Volume $= \pi r^2 h/3$

Area of curved surface $= \pi rl$

Sphere   Volume $= 4\pi r^3/3$

Surface area $= 4\pi r^2$

Trapezoidal rule

$$\text{Area} = (\text{interval width}) \times \begin{pmatrix} \text{mean of first} & \text{sum of} \\ \text{and final} & + \text{remaining} \\ \text{ordinates} & \text{ordinates} \end{pmatrix}$$

Mid-ordinate rule

$$\text{Area} = \text{interval width} \times \text{sum of mid-ordinates}$$

Simpson's rule
For an *even number* of strips of width h

$$\text{Area} = \frac{h}{3} \times \left\{ \begin{pmatrix} \text{sum of first} \\ \text{and final} \\ \text{ordinates} \end{pmatrix} + 4 \begin{pmatrix} \text{sum of} \\ \text{even} \\ \text{ordinates} \end{pmatrix} + 2 \begin{pmatrix} \text{sum of} \\ \text{remaining odd} \\ \text{ordinates} \end{pmatrix} \right\}$$

Powers or indices
a^n is the continued product of n terms of value a.

$$a^{1/n} = \sqrt[n]{a} \qquad a^{-n} = \frac{1}{a^n}$$

$$a^m \times a^n = a^{(m+n)} \qquad (a^n)^m = a^{nm}$$

$$\frac{a^n}{a^m} = a^{(n-m)} \qquad a^{n/m} = \sqrt[m]{(a^n)}$$

Note: $a^0 = 1$ for all values of a.

Logarithms

The logarithm of a number is the power to which the base must be raised in order to give the number.

$$m = \log_a n \quad \text{means that} \quad n = a^m$$

$$\log_a (mn) = \log_a m + \log_a n$$

$$\log_a \frac{m}{n} = \log_a m - \log_a n$$

$$\log_a (m^n) = n \log_a m$$

$$\log_a \sqrt[n]{m} = \frac{1}{n} \log_a m$$

$$\log_a m = \log_b m \times \log_a b = \frac{\log_b m}{\log_b a}$$

The logarithm of x to base e (natural logarithm) is written ln x. The logarithm of x to base 10 is written log x or lg x.

Powers

$$(x + a)^n = A_0 x^n + A_1 x^{n-1} a + A_2 x^{n-2} a^2 +$$

$$\cdots + A_{n-1} x a^{n-1} + A_n a^n$$

where the coefficients A_0, A_1, A_2, etc are given below. In the case of the expression $(x \pm a)^n$, the even numbered coefficients (A_0, A_2, etc) are positive, and the odd numbered coefficients are negative if $(x - a)^n$, is considered.

Value of n	Value of the coefficients A_0, A_1, etc.
1	1 1
2	1 2 1
3	1 3 3 1
4	1 4 6 4 1

Each value of A is the sum of the two values on the line above it which are immediately to the right and left of it.

$$(x \pm a)^1 = x \pm a$$

$$(x \pm a)^2 = x^2 \pm 2ax + a^2$$

$$(x \pm a)^3 = x^3 \pm 3x^2a + 3xa^2 \pm a^3$$

Completing the square
To complete the square, add the square of half the coefficient of x.

$$(x^2 \pm 2ax) + a^2 = (x \pm a)^2$$

Quadratic equations
The roots of $ax^2 + bx + c = 0$ are

$$x = \frac{-b \pm (b^2 - 4ac)^{1/2}}{2a}$$

in which $(b^2 - 4ac)$ is known as the *discriminant*.

If the discriminant is zero, the roots are real and equal.

If the discriminant is positive, the roots are real and unequal.

If the discriminant is negative, the roots are complex and unequal.

SERIES

Arithmetic progression
If A, B, C, are any three of a series of numbers in an arithmetic progression (A.P.), then

$$\text{common difference} = d = B - A = C - B$$

The nth term in the series is $A + (n - 1)d$. The sum S_n of the first n terms in the series is

$$S_n = A + (A + d) + (A + 2d) + \cdots + \{A + (n - 1)d\}$$

$$= \frac{n}{2}\{2A + (n - 1)d\}$$

Geometric progression
If A, B, C, are any three of a series of numbers in a geometric progression (G.P.), then

$$\text{common ratio} = r = \frac{B}{A} = \frac{C}{B}$$

The nth term in the series is Ar^{n-1}. The sum S_n of the first n terms in the series is

$$S_n = A + Ar + Ar^2 + \cdots + Ar^{n-1}$$

$$= \frac{A(1 - r^n)}{1 - r} \qquad \text{(N.B. } r \neq 1\text{)}$$

If r lies in the range -1 to $+1$, the sum to infinity is $A/(1 - r)$.

Mean values
The mean values of two quantities a and b are

$$\text{Arithmetic mean} = \frac{a + b}{2} = A$$

$$\text{Geometric mean} = (ab)^{1/2} = G$$

$$\text{Harmonic mean} = \frac{2ab}{a + b} = H = \frac{G^2}{A}$$

Convergency
If S_n is the sum of the first n terms in a series, then the series is *convergent* if $\text{Lim}_{n\to\infty} S_n$ is finite and unique.

Binomial series
If n is a positive integer, then the $(r + 1)$th term in the series $(x + a)^n$ is

$$\frac{n(n - 1)(n - 2) \cdots (n - r + 1)}{r!} x^{n-r} a^r$$

and the series is

$$(a + x)^n = a^n + na^{n-1}x + \frac{n(n-1)}{2!}a^{n-2}x^2$$

$$+ \frac{n(n-1)(n-2)}{3!}a^{n-3}x^3 + \cdots$$

$$+ \frac{n(n-1)(n-2)\cdots(n-r+1)}{r!}a^{n-r}x^r$$

$$+ \cdots + a^n$$

If n is not a positive integer, the series is infinite and converge only if $|x/a| < 1$.

If $n \ll 1$, then

$$(1 \pm x)^n \approx 1 \pm nx$$

If n is fractional or negative, and if $|x| < 1$, then

$$(1 \pm x)^n = 1 \pm nx + \frac{n(n-1)}{2!}x^2 \pm \frac{n(n-1)(n-2)}{3!}x^3 + \cdots$$

Exponential series

$$e^{\pm x} = 1 \pm x + \frac{x^2}{2!} \pm \frac{x^3}{3!} + \cdots \quad \text{to infinity}$$

If $x = +1$, then $e = 2.71828\ldots$

$$e^{\pm kx} = 1 \pm kx + \frac{k^2 x^2}{2!} \pm \frac{k^3 x^3}{3!} + \cdots \quad \text{to infinity}$$

Trigonometrical series

$$\sin x = x - \frac{x^3}{3!} + \frac{x^5}{5!} - \cdots \quad \text{to infinity}$$

$$\cos x = 1 - \frac{x^2}{2!} + \frac{x^4}{4!} - \cdots \quad \text{to infinity}$$

$$\tan x = x + \frac{x^3}{3} + \frac{2x^5}{15} + \frac{17x^7}{315} + \cdots \quad \text{to infinity}$$

Hyperbolic series

$$\sinh x = x + \frac{x^3}{3!} + \frac{x^5}{5!} + \cdots \quad \text{to infinity}$$

$$\cosh x = 1 + \frac{x^2}{2!} + \frac{x^4}{4!} + \cdots \quad \text{to infinity}$$

$$\tanh x = x - \frac{x^3}{3} + \frac{2x^5}{15} - \frac{17x^7}{315} + \cdots \quad \text{to infinity}$$

Logarithmic series

$$\ln (1 + x) = x - \frac{x^2}{2} + \frac{x^3}{3} - \cdots \quad (-1 < x \leq 1)$$

$$\ln (1 - x) = -x - \frac{x^2}{2} - \frac{x^3}{3} - \cdots \quad (-1 \leq x < 1)$$

Taylor's series

$$f(x + h) = f(x) + hf'(x) + \frac{h^2}{2!} f''(x) + \cdots$$

$$+ \frac{h^n}{n!} f^{(n)}(x) + \cdots \text{ to infinity}$$

where

$$f^{(n)}(x) = \frac{d^n}{dx^n} f(x)$$

and is the nth differential coefficient of $f(x)$ with respect to x.

MacLaurin's series

$$f(x) = f(0) + xf'(0) + \frac{x^2}{2!} f''(0) + \cdots$$

$$+ \frac{x^n}{n!} f^{(n)}(0) + \cdots \text{ to infinity}$$

Fourier series
For a period 2π over the range α to β (that is, $\beta - \alpha = 2\pi$)

$$f(x) = A_0 + a_1 \cos x + a_2 \cos 2x + \cdots + a_n \cos nx + \cdots$$

$$+ b_1 \sin x + b_2 \sin 2x + \cdots + b_n \sin nx + \cdots$$

$$\text{to infinity}$$

where $n = 1, 2, 3, \cdots$

$$A_0 = \frac{1}{2\pi} \int_\alpha^\beta f(x)\ dx$$

$$a_n = \frac{1}{\pi} \int_\alpha^\beta f(x) \cos nx\ dx$$

$$b_n = \frac{1}{\pi} \int_\alpha^\beta f(x) \sin nx\ dx$$

Even functions (symmetry about the y axis – see Figure 29)

$f(x) = f(-x)$ No sine terms in the Fourier series.

$f(x) = A_0 + a_1 \cos x + a_2 \cos 2x + \cdots$

A_0 may exist

a_n exists

$b_n = 0$

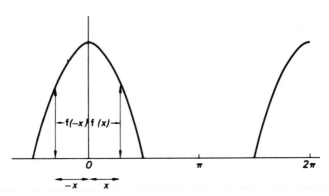

Figure 29

93

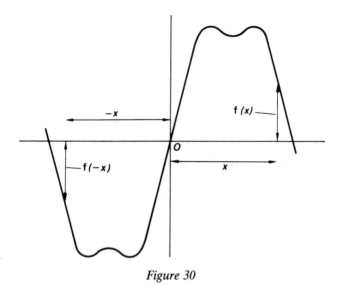

Figure 30

Odd functions (symmetry about the origin – see Figure 30)

$f(x) = -f(-x)$ No cosine terms in the Fourier terms

$f(x) = b_1 \sin x + b_2 \sin 2x + \cdots$

$A_0 = 0$

$a_n = 0$

b_n exists

Half wave repetition (see Figure 31)

$f(x) = f(x + \pi)$ No odd terms in the Fourier series.

A_0 may exist

$\left.\begin{array}{l} a_n \\ b_n \end{array}\right\}$ even terms only

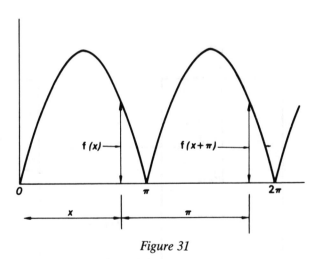

Figure 31

Half wave inversion (see Figure 32)

$f(x) = -f(x + \pi)$ No even terms in the Fourier series.

$A_0 = 0$

$\left.\begin{array}{l} a_n \\ b_n \end{array}\right\}$ odd terms only

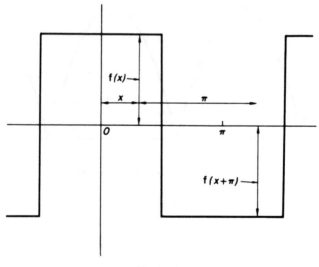

Figure 32

TRIGONOMETRY

The *radian* is the angle subtended at the centre of any circle by an arc equal in length to the radius.

2π radians correspond to 360°
1 radian corresponds to 57.3°

Basic equations
If a, b and c are the sides of a right-angled triangle (see Figure 33) and c is the *hypotenuse* then, with respect to angle α, a is called the *opposite side*, and b the *adjacent side*.

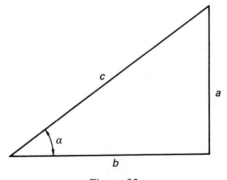

Figure 33

$$c^2 = a^2 + b^2$$

$$\sin \alpha = \frac{a}{c} \qquad\qquad \operatorname{cosec} \alpha = \frac{1}{\sin \alpha} = \frac{c}{a}$$

$$\cos \alpha = \frac{b}{c} \qquad \sec \alpha = \frac{1}{\cos \alpha} = \frac{c}{b}$$

$$\tan \alpha = \frac{a}{b} \qquad \cot \alpha = \frac{1}{\tan \alpha} = \frac{b}{a}$$

$\sin^{-1} x$ (arcsin x) is the angle whose sine is x
$\cos^{-1} x$ (arccos x) is the angle whose cosine is x
$\tan^{-1} x$ (arctan x) is the angle whose tangent is x
$\sin (-\alpha) = -\sin \alpha$
$\cos (-\alpha) = \cos \alpha$
$\tan (-\alpha) = -\tan \alpha$
$\sin (90° - \alpha) = \sin (90° + \alpha) = \cos \alpha$
$\cos (90° - \alpha) = -\cos (90° + \alpha) = \sin \alpha$
$\tan (90° - \alpha) = -\tan (90° + \alpha) = \cot \alpha$

Fundamental relationships

$$\sin^2 \alpha + \cos^2 \alpha = 1$$

$$1 + \tan^2 \alpha = \sec^2 \alpha$$

$$1 + \cot^2 \alpha = \csc^2 \alpha$$

$$\tan \alpha = \frac{\sin \alpha}{\cos \alpha}$$

Formulae for the solution of triangles
In the following a, b, and c are the sides of a triangle, angle A being opposite side a, angle B being opposite side b, and angle C being opposiste side c. The radius of the circumscribing circle passing through the three vertices is R.

$$\frac{a}{\sin A} = \frac{b}{\sin B} = \frac{c}{\sin C} = 2R$$

$$a^2 = b^2 + c^2 - 2bc \cos A$$

$$b^2 = a^2 + c^2 - 2ac \cos B$$

$$c^2 = a^2 + b^2 - 2ab \cos C$$

Compound angle formulae

$$\sin (A \pm B) = \sin A \cos B \pm \cos A \sin B$$

$$\cos (A \pm B) = \cos A \cos B \mp \sin A \sin B$$

$$\tan (A \pm B) = \frac{\tan A \pm \tan B}{1 \mp \tan A \tan B}$$

$$\sin 2A = 2 \sin A \cos A$$

$$\cos 2A = \cos^2 A - \sin^2 A = 2 \cos^2 A - 1 = 1 - 2 \sin^2 A$$

$$2 \sin A \cos B = \sin(A + B) + \sin(A - B)$$

$$2 \cos A \sin B = \sin(A + B) - \sin(A - B)$$

$$2 \cos A \cos B = \cos(A + B) + \cos(A - B)$$

$$2 \sin A \sin B = \cos(A - B) - \cos(A + B)$$

$$\sin A + \sin B = 2 \sin \frac{A - B}{2} \cos \frac{A + B}{2}$$

$$\sin A - \sin B = 2 \cos \frac{A + B}{2} \sin \frac{A - B}{2}$$

$$\cos A + \cos B = 2 \cos \frac{A + B}{2} \cos \frac{A - B}{2}$$

$$\cos A - \cos B = -2 \sin \frac{A+B}{2} \sin \frac{A-B}{2}$$

$$A \cos \theta + B \sin \theta = \sqrt{(A^2 + B^2)} \cos (\theta + \tan^{-1}(-B/A))$$

Hyperbolic forms of trigonometrical functions

$$\sin \alpha = \frac{1}{2j} (e^{j\alpha} - e^{-j\alpha}) \qquad \cos \alpha = \frac{1}{2} (e^{j\alpha} + e^{-j\alpha})$$

$$\sinh \alpha = \frac{1}{2} (e^{\alpha} - e^{-\alpha}) \qquad \cosh \alpha = \frac{1}{2} (e^{\alpha} + e^{-\alpha})$$

$$j \sin \alpha = \sinh j\alpha \qquad \sin j\alpha = j \sinh \alpha$$

$$\cos \alpha = \cosh j\alpha \qquad \cos j\alpha = \cosh \alpha$$

$$j \tan \alpha = \tanh j\alpha \qquad \tan j\alpha = j \tanh \alpha$$

Hyperbolic formulae

$$\tanh A = \sinh A / \cosh A$$

$$\coth A = \cosh A / \sinh A$$

$$\operatorname{sech} A = 1 / \cosh A$$

$$\operatorname{cosech} A = 1 / \sinh A$$

$$\cosh^2 A - \sinh^2 A = 1$$

$$1 - \tanh^2 A = \operatorname{sech}^2 A$$

$$1 - \coth^2 A = -\operatorname{cosech}^2 A$$

$$\cosh(-A) = \cosh A$$

$$\sinh(-A) = -\sinh A$$

DETERMINANTS

The solution of the simultaneous equations

$$a_1x + b_1y = A$$
$$a_2x + b_2y = B$$

is given by

$$\frac{x}{\Delta_x} = \frac{y}{\Delta_y} = \frac{1}{\Delta}$$

where

$$\Delta_x = \begin{vmatrix} A & b_1 \\ B & b_2 \end{vmatrix} = Ab_2 - Bb_1$$

$$\Delta_y = \begin{vmatrix} a_1 & A \\ a_2 & B \end{vmatrix} = a_1B - a_2A$$

$$\Delta = \begin{vmatrix} a_1 & b_1 \\ a_2 & b_2 \end{vmatrix} = a_1b_2 - a_2b_1$$

The solution of the simultaneous equations

$$a_1x + b_1y + c_1z = A$$
$$a_2x + b_2y + c_2z = B$$
$$a_3x + b_3y + c_3z = C$$

is given by

$$\frac{x}{\Delta_x} = \frac{y}{\Delta_y} = \frac{z}{\Delta_z} = \frac{1}{\Delta}$$

where

$$\Delta_x = \begin{vmatrix} A & b_1 & c_1 \\ B & b_2 & c_2 \\ C & b_3 & c_3 \end{vmatrix} = A\begin{vmatrix} b_2 & c_2 \\ b_3 & c_3 \end{vmatrix} - b_1\begin{vmatrix} B & c_2 \\ C & c_3 \end{vmatrix} + c_1\begin{vmatrix} B & b_2 \\ C & b_3 \end{vmatrix}$$

$$= A(b_2c_3 - b_3c_2) - b_1(Bc_3 - Cc_2) + c_1(Bb_3 - Cb_2)$$

$$\Delta_y = \begin{vmatrix} a_1 & A & c_1 \\ a_2 & B & c_2 \\ a_3 & C & c_3 \end{vmatrix}$$

$$= a_1(Bc_3 - Cc_2) - A(a_2c_3 - a_3c_2) + c_1(a_2C - a_3B)$$

$$\Delta_z = \begin{vmatrix} a_1 & b_1 & A \\ a_2 & b_2 & B \\ a_3 & b_3 & C \end{vmatrix}$$

$$= a_1(b_2C - b_3B) - b_1(a_2C - a_3B) + A(a_2b_3 - a_3b_2)$$

$$\Delta = \begin{vmatrix} a_1 & b_1 & c_1 \\ a_2 & b_2 & c_2 \\ a_3 & b_3 & c_3 \end{vmatrix}$$

$$= a_1(b_2c_3 - b_3c_2) - b_1(a_2c_3 - a_3c_2) + c_1(a_2b_3 - a_3b_2)$$

MATRICES

Matrix representation

The equations

$$V_1 = AI_1 + BI_2 + CI_3$$
$$V_2 = DI_1 + EI_2 + FI_3$$
$$V_3 = GI_1 + HI_2 + JI_3$$

can be written in *matrix form* of ordered rows and columns of elements as follows

$$\begin{bmatrix} V_1 \\ V_2 \\ V_3 \end{bmatrix} = \begin{bmatrix} A & B & C \\ D & E & F \\ G & H & J \end{bmatrix} \begin{bmatrix} I_1 \\ I_2 \\ I_3 \end{bmatrix}$$

where V_1, V_2 and V_3 are voltages forming a *column matrix*; I_1, I_2 and I_3 are currents forming another column matrix; and A, B, C, D, E, F, G, H and J are resistance (or impedance) values which form a *square matrix*.

A *rectangular matrix* having M rows and N columns ($M = N$) is described as an M *by* N *matrix* or an $M \times N$ *matrix*.

Matrix addition

Two matrices can be *added together* or *subtracted from one another* if they are of the same order (that is, they must

103

both have the same number of rows and the same number of columns). In this case

$$A \pm B = [a_{ij} + b_{ij}]$$

where a_{ij} and b_{ij} are, respectively, the element in the ith row and j^{th} column of matrices A and B.

$$\text{If } A = \begin{bmatrix} 4 & 5 \\ -6 & 7 \end{bmatrix} \text{ and } B = \begin{bmatrix} 4 & -5 \\ -7 & 8 \end{bmatrix}$$

$$\text{then } A + B = \begin{bmatrix} 4+4 & 5-5 \\ -6-7 & 7+8 \end{bmatrix} = \begin{bmatrix} 8 & 0 \\ -13 & 15 \end{bmatrix}$$

$$\text{and } A - B = \begin{bmatrix} 4-4 & 5-(-5) \\ -6-(-7) & 7-8 \end{bmatrix} = \begin{bmatrix} 0 & 10 \\ 1 & -1 \end{bmatrix}$$

Matrix multiplication

Two matrices can be *multiplied together* if the number of columns in A is equal to the number of rows in B. *The product must be carried out in the order AB* (matrix B may not necessarily conform to A for multiplication; i.e., the product BA may not be allowed). The following should be observed.

1. $AB = BA$ generally.
2. $AB = 0$ does not always imply that $A = 0$ or $B = 0$.
3. $AB = AC$ does not always imply $B = C$.

Matrix multiplication is performed on a *row by column basis*; each element in a row is multiplied by the corresponding element of a column, and the products are summed.

If $A = [-1\ 2\ 3]$ and $B = \begin{bmatrix} -6 \\ 0 \\ 7 \end{bmatrix}$ then

$$AB = [-1\ 2\ 3] \begin{bmatrix} -6 \\ 0 \\ 7 \end{bmatrix} = [-1(-6) + 2(0) + 3(7)] = [27]$$

DIFFERENTIAL CALCULUS

The *differential coefficient* (the *derivative*) of y with respect to x may be expressed in several ways, including

$$\frac{dy}{dx},\ \frac{d}{dx}\,y,\ Dy,\ D_x y,\ y'$$

y	$\dfrac{dy}{dx}$
ax^n	anx^{n-1}
ae^{kx}	ake^{kx}
$\sin(ax + b)$	$a\cos(ax + b)$
$\cos(ax + b)$	$-a\sin(ax + b)$
$\tan x$	$\sec^2 x$
$\cot x$	$-\operatorname{cosec}^2 x$
$\ln x$	$1/x$
$\ln(ax + b)$	$\dfrac{a}{ax + b}$
$\sin^{-1}\dfrac{x}{a}$	$1/(a^2 - x^2)^{1/2}$

y	$\dfrac{dy}{dx}$
$\cos^{-1}\dfrac{x}{a}$	$-1/(a^2 - x^2)^{1/2}$
$\tan^{-1}\dfrac{x}{a}$	$a/(x^2 + a^2)$
$\sinh x$	$\cosh x$
$\cosh x$	$\sinh x$
$\tanh x$	$\text{sech}^2\, x$
$\sec ax$	$a \sec ax.\tan ax$
$\text{cosec}\, ax$	$-a \,\text{cosec}\, ax.\cot ax$
$\text{sech}\, ax$	$-a\, \text{sech}\, ax.\coth ax$
$\text{cosech}\, ax$	$-a\, \text{cosech}\, ax.\coth ax$
$\coth ax$	$-a\, \text{cosech}^2 ax$
$\sinh^{-1} ax$	$1/\sqrt{(x^2 + a^2)}$
$\cosh^{-1}(x/a)$	$1/\sqrt{(x^2 - a^2)}$
$\tanh^{-1}(x/a)$	$a/(a^2 - x^2)$

Product and quotient rules
$$\frac{d}{dx}(uv) = u\,\frac{dv}{dx} + v\,\frac{du}{dx}$$

$$\frac{d}{dx}\left(\frac{u}{v}\right) = \frac{v\,\dfrac{du}{dx} - u\,\dfrac{dv}{dx}}{v^2}$$

Leibnitz theorem
Allowing $u^{(p)}$ and $v^{(p)}$ to denote the pth derivatives of u and v with respect to x, then

$$\frac{d^n}{dx^n}(uv) = u^{(n)}v + nu^{(n-1)}v^{(1)} + \frac{n(n-1)}{2!}u^{(n-2)}v^{(2)}$$

$$+\frac{n(n-1)(n-2)}{3!}u^{(n-3)}v^{(3)} + \ldots + nu^{(1)}v^{(n-1)} + uv^{(n)}$$

Function of a function
If $y = f(u)$, where $u = \phi(x)$, then

$$\frac{dy}{dx} = \frac{dy}{du} \cdot \frac{du}{dx}$$

Maximum and minimum values of f(x)
A curve has a *maximum value* at a point if, at that point,

$\dfrac{dy}{dx} = 0$ and $\dfrac{d^2y}{dx^2}$ has a negative value.

A curve has a *minimum value* at a point if, at that point

$\dfrac{dy}{dx} = 0$ and $\dfrac{d^2y}{dx^2}$ has a positive value.

If at a point p on a curve $\dfrac{dy}{dx} = 0$ and $\dfrac{d^2y}{dx^2} = 0$, the curve

may either have a maximum point, or a minimum point,

or a *point of inflection*. If $\dfrac{d^2y}{dx^2}$ changes sign as x changes

from $(p - \delta x)$ to $(p + \delta x)$, then there is a point of inflection at p.

The total differential
If $z = f(x, y)$, and both x and y are independent variables, then

$$dz = \frac{\partial z}{\partial x} \, dx + \frac{\partial z}{\partial y} \, dy$$

107

INTEGRAL CALCULUS

Standard integrals

y	$\int y \, dx$
kx^n	$\dfrac{kx^{n+1}}{n+1}$ $(n \neq 1)$
ke^{ax}	$\dfrac{k}{a} e^{ax}$
$\sin(ax+b)$	$-\dfrac{1}{a}\cos(ax+b)$
$\cos(ax+b)$	$\dfrac{1}{a}\sin(ax+b)$
$\dfrac{1}{x}$	$\ln x$
$\dfrac{1}{ax+b}$	$\dfrac{1}{a}\ln(ax+b)$
$\sec^2 x$	$\tan x$
$\operatorname{cosec}^2 x$	$-\cot x$
$\dfrac{1}{(a^2-x^2)^{1/2}}$	$\sin^{-1}\dfrac{x}{a}$
$\dfrac{1}{x^2+a^2}$	$\dfrac{1}{a}\tan^{-1}\dfrac{x}{a}$
$\cosh x$	$\sinh x$
$\sinh x$	$\cosh x$
$\operatorname{sech}^2 x$	$\tanh x$

Integration by parts

$$\int u \, \frac{dv}{dx} \, dx = uv - \int v \, \frac{du}{dx} \, dx$$

108

OPERATOR D

$$Dy = \frac{dy}{dx}$$

$$D^{(n)}y = \frac{d^n y}{dx^n}$$

$$\frac{1}{D} f(x) = D^{-1}\{f(x)\} = \int f(x)\, dx$$

If $f(D) = a_0 D^{(n)} + a_1 D^{(n-1)} + \ldots + a_r D^{(n-r)} + \ldots + a_n$
where a_r is a constant and n is a positive integer, then

$$f(D)e^{ax} = e^{ax} f(a)$$

$$\frac{1}{f(D)} e^{ax} = \frac{1}{f(a)} e^{ax} \qquad [f(a) \neq 0]$$

$$f(D)\{e^{ax}f(x)\} = e^{ax}f(D + a)f(x)$$

$$\frac{1}{f(D)}\{e^{ax}f(x)\} = e^{ax} \frac{1}{f(D + a)} f(x)$$

$$\frac{1}{f(D^2)} \sin ax = \frac{1}{f(-a^2)} \sin ax$$

$$\frac{1}{f(D^2)} \cos ax = \frac{1}{f(-a^2)} \cos ax$$

$$[f(a) \neq 0]$$

PARTIAL FRACTIONS

If $f(x) = \dfrac{A(x)}{M(x)}$, where $A(x)$ and $M(x)$ are polynomials in x, the degree of $A(x)$ being less than that of $M(x)$, then

1. For every linear factor $(ax + b)$ of $M(x)$ there is a corresponding partial fraction $\dfrac{1}{ax + b}$.

2. For every quadratic factor $(ax^2 + bx + c)$ of $M(x)$ there is a corresponding partial fraction $\dfrac{px + q}{ax^2 + bx + c}$.

3. For every repeated factor $(ax + b)^2$ of $M(x)$ there is a corresponding partial fraction $\dfrac{p}{ax + b} + \dfrac{q}{(ax + b)^2}$

4. For every repeated quadratic factor $(ax^2 + bx + c)^2$ in $M(x)$ there is a corresponding partial fraction

$$\dfrac{px + q}{ax^2 + bx + c} + \dfrac{rx + s}{(ax^2 + bx + c)^2}$$

'Cover-up' rule
If

$$f(x) = \dfrac{p(x)}{(x + a)(x + b)(x + c) \cdots}$$

then the numerators of the separate fractions due to the factors $(x + a)$, $(x + b)$, etc, are determined by 'covering up' each of the factors in turn and evaluating the remainder of the expression by replacing each x term by the value of x which makes the 'covered up' factor zero.

For example, if

$$f(x) = \frac{x^2 + x + 2}{(x + 3)(x + 2)(x + 1)}$$

then

$$f(x) = \frac{\{(-3)^2 - 3 + 2\}/(-3 + 2)(-3 + 1)}{x + 3}$$

$$+ \frac{\{(-2)^2 - 2 + 2\}/(-2 + 3)(-2 + 1)}{x + 2}$$

$$+ \frac{\{(-1)^2 - 1 + 2\}/(-1 + 3)(-1 + 2)}{x + 1}$$

$$= \frac{4}{x + 3} - \frac{4}{x + 2} + \frac{1}{x + 1}$$

LAPLACE TRANSFORMS

If F(t) is a function of t for values of $t > 0$, then the Laplace transformation of F(t) is

$$\int_0^\infty e^{-st} F(t) dt$$

F(t)	Laplace transformation
1 (unit step function)	$1/s$
A	A/s
δ (unit impulse function)	1
$e^{-\alpha t}$	$\dfrac{1}{s + \alpha}$
$1 - e^{-\alpha t}$	$\dfrac{\alpha}{s(s + \alpha)}$
$e^{-\alpha t} - e^{-\beta t}$	$\dfrac{\beta - \alpha}{(s + \alpha)(s + \beta)}$
t (ramp function)	$\dfrac{1}{s^2}$
$te^{-\alpha t}$	$\dfrac{1}{(s + \alpha^2)}$
t^n (n a positive integer)	$\dfrac{n!}{s^{n + 1}}$
$t^n e^{-\alpha t}$	$\dfrac{n!}{(s + \alpha)^{n + 1}}$

F(t)	Laplace transformation
$\sin \omega t$	$\dfrac{\omega}{s^2 + \omega^2}$
$\cos \omega t$	$\dfrac{s}{s^2 + \omega^2}$
$\sin(\omega t + \phi)$	$\dfrac{s \sin \phi + \omega \cos \phi}{s^2 + \omega^2}$
$e^{-\alpha t} \sin \omega t$	$\dfrac{\omega}{(s + \alpha)^2 + \omega^2}$
$e^{-\alpha t} \cos \omega t$	$\dfrac{s + \alpha}{(s + \alpha)^2 + \omega^2}$
$\sinh \alpha t$	$\dfrac{\alpha}{s^2 - \alpha^2}$
$\cosh \alpha t$	$\dfrac{s}{s^2 - \alpha^2}$
$e^{-\beta t} \sinh \alpha t$	$\dfrac{\alpha}{(s + \beta)^2 - \alpha^2}$
$e^{-\beta t} \cosh \alpha t$	$\dfrac{s + \beta}{(s + \beta)^2 - \alpha^2}$
Delayed step function	$e^{-sT/s}$
Rectangular pulse	$(1 - e^{-sT})/s$
$\displaystyle\int_0^t f(t)\, dt$	$F(s)/s$
$e^{-\alpha t} \left(\cos \omega t - \dfrac{\alpha}{\omega} \sin \omega t \right)$	$\dfrac{s}{(s + \alpha)^2 + \omega^2}$

F(t)	Laplace transformation
$\sin(\omega t \pm \phi)$	$\dfrac{\omega \cos \phi \pm s \sin \phi}{s^2 + \omega^2}$
$\cos(\omega t \pm \phi)$	$\dfrac{s \cos \phi \pm \omega \sin \phi}{s^2 + \omega^2}$
$t \sin \omega t$	$\dfrac{2\omega s}{(s^2 + \omega^2)^2}$
$t \cos \omega t$	$\dfrac{s^2 - \omega^2}{(s^2 + \omega^2)^2}$

If F(s) is the Laplace transform of F(t), then

$$L\{F'(t)\} = sF(s) - F(0)$$

$$L\{F''(t)\} = s^2F(s) - sF(0) - F'(0)$$

$$L\{F^{(n)}(t)\} = s^nF(s) - s^{n-1}F(0) - s^{n-2}F'(0) - \cdots$$
$$- sF^{(n-2)}(0) - F^{(n-1)}(0)$$

where $F^{(n-r)}(0)$ is the value of the rth differential of F(t) when $t \to 0$.

Initial value theorem

$$f(0^+) = \lim_{s \to \infty} s\, F(s)$$

Final value theorem

$$f(\infty) = \lim_{s \to 0} s\, F(s)$$

Note: When applying the final value theorem, *beware of poles on the jω axis*.

z-TRANSFORMS

$$z = e^{Ts}$$

where s is the Laplace Transform of the variable, and T is the sampling period. The z-transform is defined by

$$Z\{f(n)\} = F(z) = \sum_{n=0}^{\infty} f(n) \, z^{-n}$$

where $f(n)$ is the sampled version of $f(t)$, and $n = 1, 2, 3 \ldots$ refers to discrete sampling times $t_1, t_2, t_3 \ldots$.

F(t)	z-Transform
1 (unit step function)	$z/(z - 1)$
δ (unit impulse function)	1
t (ramp function)	$\dfrac{Tz}{(z - 1)^2}$
$\dfrac{t^2}{2}$	$\dfrac{T^2 z(z + 1)}{2(z - 1)^3}$

115

F(t)	z-Transform
$e^{-\alpha t}$	$\dfrac{z}{z - e^{-\alpha T}}$
$te^{-\alpha t}$	$\dfrac{Tze^{-\alpha T}}{(z - e^{-\alpha T})^2}$
$1 - e^{-\alpha t}$	$\dfrac{z(1 - e^{-\alpha T})}{(z - 1)(z - e^{-\alpha T})}$
$\sin \omega t$	$\dfrac{z \sin \omega T}{z^2 - 2z \cos \omega T + 1}$
$e^{-\alpha t} \sin \omega t$	$\dfrac{ze^{-\alpha T} \sin \omega T}{z^2 e^{2\alpha T} - 2ze^{\alpha T} \cos \omega T + 1}$
$\cos \omega t$	$\dfrac{z(z - \cos \omega T)}{z^2 - 2z \cos \omega T + 1}$
$e^{-\alpha T} \cos \omega T$	$\dfrac{z^2 - ze^{-\alpha T} \cos \omega T}{z^2 - 2ze^{-\alpha T} \cos \omega T + e^{-2\alpha T}}$

Initial value theorem

$$f(0^+) = \lim_{z \to \infty} s\, F(z)$$

Final value theorem

$$\lim_{n \to \infty} f(n) = \lim_{z \to 1} (1 - z^{-1})\, F(z)$$

F(z) *must not* have any poles *on or outside the unit circle*.